# LOW-COST SPACE-BORNE DATA FOR INUNDATION MODELLING: TOPOGRAPHY, FLOOD EXTENT AND WATER LEVEL

# Low-cost space-borne data for inundation modelling: topography, flood extent and water level

DISSERTATION

Submitted in fulfilment of the requirements of
the Board for Doctorate of Delft University of Technology
and of
the Academic Board of the UNESCO-IHE Institute for Water Education
for the Degree of DOCTOR
to be defended in public
on Thursday, 9 July 2015 at 12:30 hours
in Delft, the Netherlands

by

**Kun YAN**

Master of Science Water Science and Engineering
specialization in Hydroinformatics
UNESCO-IHE, Institute for Water Education

born in Bengbu, Anhui, China

This dissertation has been approved by the promotor:
Prof. dr. D.P. Solomatine

Composition of Doctoral Committee:

| | |
|---|---|
| Chairman | Rector Magnificus TU Delft |
| Vice-Chairman | Rector UNESCO-IHE |
| Prof. dr. D.P. Solomatine | UNESCO-IHE/TU Delft, promotor |
| Prof. dr. G. Di Baldassarre | Uppsala University, Sweden |

Independent Members:
| | |
|---|---|
| Prof. dr. ir. P. Willems | KU Leuven, Belgium |
| Prof. dr. ir. A.B.K. van Griensven | Vrije Universiteit Brussel, Belgium /UNESCO-IHE |
| Prof. dr. ir. W.G.M. Bastiaanssen | TU Delft/UNESCO-IHE |
| Prof. dr. ir. A.E. Mynett | UNESCO-IHE/TU Delft |
| Prof.dr.ir.W.S.J. Uijttewaal | TU Delft, reserve member |

Prof. dr. G. Di Baldassarre, Uppsala University, Sweden, has, as supervisor, contributed significantly to the preparation of this dissertation.

CRC Press/Balkema is an imprint of the Taylor & Francis Group, an informa business

Published by:
CRC Press/Balkema
PO Box 11320, 2301 EH Leiden, The Netherlands
e-mail: Pub.NL@taylorandfrancis.com
www.crcpress.com – www.taylorandfrancis.com
ISBN: 978-1-138-02875-3 (Taylor & Francis Group)

# SUMMARY

Floods are among the most damaging natural hazards and their impacts have been dramatically increasing worldwide over the past decades. As most basins of the world are ungauged or poorly gauged and some measurement networks are continuously under decline, the spatial distribution of flood hazard is often difficult to estimate because the input data needed for flood inundation modelling (e.g. topographies, flood extents, water levels) are often not available.

A unique opportunity is nowadays provided by the ongoing development of remote sensing data, such as the low-cost, space-borne data. In particular, the development of new remotely sensed data sources has not only shifted flood modelling from a data-poor to a data-rich environment, but also provided a paradigm shift in flood modelling: from developing more sophisticated flood models to evaluating potential of remote sensing data. There is a general consensus that the increased availability and quality of those low-cost remote sensing data will be valuable for improving prediction in ungauged basins. However, their value and potential in supporting hydraulic modelling of floods are still not sufficiently explored in view of the unavoidable, intrinsic uncertainty affecting any modeling exercise. In this context, this thesis aims to explore the potential and limitations of low-cost, space-borne data in flood inundation modelling under uncertainty.

In our research work, we analyze the potential in supporting hydraulic modelling of floods of: NASA's SRTM (Shuttle Radar Topographic Mission) topographic data, SAR (Synthetic Aperture Radar) satellite imagery and radar altimetry. The characteristics of those data, and their pros and cons for inundation modelling are discussed. For example, SRTM`s global coverage and relatively low vertical error on low-slope areas are in favour of floodplain modelling, while its absence of in-channel geometry information would hamper its application in flood studies. Low-cost SAR imagery`s day-night, all-weather, cloud-free acquisition are particularly useful for flood extent monitoring, while its low resolution could induce equifinality in inundation model conditioning. Radar altimetry`s reliable water level measurements over large rivers provides opportunities for flood model calibration and evaluation, while its low space-time frequency limits the application in areas such as flood forecasting.

To this end, research work has been carried out by either following a model calibration-evaluation approaches or by explicitly considers major sources of uncertainty within a Monte Carlo framework. To generalize our findings, three river reaches with various scales (from medium to large) and topographic characteristics (e.g. valley-filling, two-level embankments, large and flat floodplain) are used as test sites. Thus, specific modelling exercises are implemented with slight, tailor-made modifications to deal with practical issues, such as the actual data availability, the characteristics of flood events etc. The usefulness of the low-cost space-borne data is quantitatively analyzed. Lastly, an application of SRTM-based flood modelling of a large river is conducted to highlight the challenges of predictions in ungauged basins.

The outcomes of the study provide indications on the potential and limitations of low-cost, space-borne data in supporting flood inundation modelling under uncertainty. Specifically, DEM resolution is often less of an issue than its vertical accuracy, as long as the coarse resolution allows the representation of flood pattern-controlling topographic features for the flood modelling issue, which is often not the case in urban flood studies. Thus, the thesis includes and discusses the usefulness of these data according to specific modelling purpose (e.g. re-insurance, planning, design). Moreover, topographic uncertainty could be compensated by other sources of uncertainties in hydraulic modelling if they are explicitly taken into account. The model prediction based on SRTM can be very close to that based on high-resolution, high-accuracy topographic data under other sources of uncertainty. However, besides modelling purpose and uncertainty considered, their actual usefulness could be affected by several other factors, such as the scale of the river under study, flood frequency, and the choice of modelling tools. Furthermore, the issue of in-channel information absent in SAR-derived DEMs are also discussed. It could be partially resolved by using either the global river depth dataset, or depth estimating from hydraulic geometry theory or model parameterization. Lastly, we discuss the upcoming satellite missions, which could potentially impact the way we model flood inundation patters.

Kun Yan

Delft, the Netherlands

# SAMENVATTING

Overstromingen behoren tot de meest schadelijke rampen en de gevolgen ervan zijn de afgelopen decennia wereldwijd dramatisch toegenomen. Aangezien de meeste stroomgebieden wereldwijd onbemeten of slecht bemeten zijn en het aantal meetsystemen achteruit blijft gaan, is de ruimtelijke spreiding van overstromingsgevaar vaak moeilijk in te schatten, omdat de invoerdata voor overstromingsmodellen (bv. topografie, overstroomd oppervlak, waterstanden) vaak niet beschikbaar zijn.

Tegenwoordig zorgt de verdere ontwikkeling van remote sensing data, zoals goedkope satelliet data, voor unieke mogelijkheden. Het gebruik van nieuwe remote sensing data heeft niet alleen het modelleren van overstromingen van gegevens arm in gegevens rijk veranderd, maar versterkt ook de paradigmaverschuiving in de modellering van overstromingen: Van het ontwikkelen van meer geavanceerde overstromingsmodellen tot het gebruik van remote sensing data in deze modellen. Er is een algemene consensus dat de toegenomen beschikbaarheid en kwaliteit van goedkope remote sensing data, waardevol zal zijn voor het verbeteren van voorspellingen in onbemeten stroomgebieden. Hun waarde en potentie in de ondersteuning van hydraulische modellering van overstromingen zijn echter nog steeds niet voldoende onderzocht, met het oog op de onvermijdelijke, intrinsieke onzekerheid van elk model. Vanuit deze context is dit proefschrift gericht op het verkennen van de mogelijkheden en beperkingen van de low-cost satelliet gegevens in overstromingsmodellen onder onzekere omstandigheden.

In ons onderzoek, analyseren we de mogelijkheid om hydraulische overstromingsmodellen te verbeteren met: NASA's SRTM (Shuttle Radar Topographic Mission) topografische gegevens, SAR (Synthetic Aperture Radar) satellietbeelden en radaraltimetrie. De kenmerken van deze gegevens, en hun voor- en nadelen voor overstromingsmodellen worden besproken. SRTM heeft een wereldwijde dekking en relatief lage verticale fout op gebieden met kleine hoogteverschillen en zijn bijvoorbeeld zeer geschikt voor het modelleren van uiterwaarden, terwijl de afwezigheid van geometrische informatie over de rivierbedding de toepassing ervan in overstroming studies zouden kunnen belemmeren. Low-cost SAR beelden die niet beïnvloed worden door het weer, bewolking of afwezigheid van daglicht, zijn vooral nuttig om de omvang van de overstromingen te monitoren. Maar de lage resolutie kan equifinality veroorzaken in overstromingsmodellen. Betrouwbare hoogtemeting van het water niveau van grote rivieren met behulp van radar biedt mogelijkheden voor kalibratie en evaluatie van overstromingsmodellen, terwijl de lage ruimte en tijd frequentie de toepassing beperkt voor het voorspellen van overstromingen.

Met dat doel is onderzoek uitgevoerd door het volgen van model kalibratie en evaluatie methodes of door een expliciete benadering van de belangrijke bronnen van onzekerheid op basis van Monte Carlo analyses. Om onze bevindingen te generaliseren zijn er drie rivier secties van verschillende grote (gemiddeld tot groot) en topografische kenmerken (vallei vullend, dubbele bedijking, grote en vlakke uiterwaarden) gebruikt als test locaties. Zo worden specifieke model tests uitgevoerd met kleine, op maat gemaakte aanpassingen, om zodoende om te gaan met praktische zaken, zoals de feitelijke beschikbaarheid van gegevens, de kenmerken van overstromingen enz. De bruikbaarheid van de low-cost, vanuit satelliet observatie verkregen, gegevens is vervolgens kwantitatief geanalyseerd. Ten slotte is een toepassing van een op SRTM gebaseerd overstromingsmodel van een grote rivier uitgevoerd om de uitdagingen van voorspellingen in onbemeten stroomgebieden te benadrukken.

De uitkomsten van het onderzoek geven een indicatie van de mogelijkheden en beperkingen van de low-cost, vanuit satelliet observatie verkregen gegevens, voor het verbeteren van overstromingsmodellen onder onzekere omstandigheden. Hierbij is de DEM resolutie vaak een kleiner probleem dan de verticale nauwkeurigheid, zolang de grove resolutie de belangrijke topografische elementen maar goed weergeeft die het overstromingspatroon kunnen beïnvloeden, wat vaak niet het geval is in studies naar overstromingen in stedelijke gebieden. Dit proefschrift bespreekt de bruikbaarheid van deze gegevens naargelang het specifieke doel van modellering (bv. herverzekering, planning, ontwerp). Bovendien zou de topografische onzekerheid kunnen worden gecompenseerd door andere bronnen van onzekerheid in de hydraulische modellen als daar expliciet rekening mee wordt gehouden. De voorspelling van de op SRTM gebaseerde modellen kunnen bijna dezelfde prestaties hebben als modellen die gebaseerd zijn op zeer nauwkeurige topografische gegevens met een hoge resolutie vanwege andere bronnen van onzekerheid. Echter, naast het gebruik voor doelen en onzekerheden in modellen, zou het werkelijke nut ervan door verscheidene andere factoren beïnvloed kunnen worden, zoals de omvang van de bestudeerde rivier, overstromingsfrequentie, en de keuze van modelleringinstrumenten. Verder wordt ook het ontbreken van informatie over het rivierbed van op SAR gebaseerde DEMs besproken. Dit kan gedeeltelijk worden opgelost met behulp van ofwel de dataset voor globale rivierdieptes, of schattingen van waterdiepte op basis van hydraulische geometriemodellen, of door model parametrisatie. Tot slot bespreken we de verwachte nieuwe satelliet missies, die mogelijk een invloed hebben op de manier waarop we overstromingspatronen modelleren.

Kun Yan

Delft, the Netherlands

# CONTENTS

# CHAPTER 1

# INTRODUCTION

This chapter introduces low-cost space-borne data such as SRTM topography, satellite imagery and radar altimetry, as well as the current progress in integrating various low-cost space-borne data for inundation modelling toward model building, calibration and evaluation, mainly in the last five years. At the end of this chapter, the objectives, methodology and outline of this thesis are presented.

## 1.1   BACKGROUND

Floods are among the most damaging natural hazards and their impacts have been dramatically increasing worldwide over the past decades (Dankers et al., 2013; Aerts et al., 2014; Di Baldassarre et al., 2010). Flood risk is likely to increase in the near future not only due to climate change and sea level rise, but also because of the growth of human population in floodplains (Figure 1.1, Di Baldassarre et al., 2010; see also Hinkel et al., 2014; Jongman et al., 2014). Anticipating flood risk in a changing environment is therefore crucial for sustainable development in the 21st century.

*Figure 1.1* Flood risk and population growth: spatial distribution of population growth in the period between 1960 and 2000 (scale from yellow to red); and location of floods (dots) and deadly floods (black circles) (redraw from Di Baldassarre et al., 2010)

Hydraulic modelling has become an essential tool for flood risk studies. However, as most basins of the world are ungauged or poorly gauged (Stokstad, 1999; Sivapalan et al., 2003) and many measurement networks are continuously under decline, there is often a lack of data for hydraulic modelling of floods. The initiative launched by the International Association of Hydrological Sciences (IAHS) on Predictions in Ungauged Basins (PUB) more than ten years ago has highlighted the need to advance predictions in ungauged basins. One of the science themes of PUB was to exploit the wealth of new data sources for improved prediction in ungauged basins or data-sparse areas. Specifically, one question has been raised in the hydrological sciences community: 'How can we employ new observational technologies in improved predictive methods?' The integration of such 'new' observations with hydrologic/hydraulic modelling has therefore been one of the focuses in hydrological sciences over the past decade (Sivapalan et al., 2003).

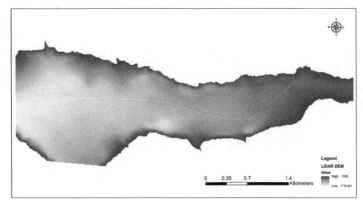

*Figure 1.2* *LiDAR data of City of Barcelonnette, France*

In this context, the development of new remotely sensed data sources has not only shifted flood modelling from a data-poor to a data-rich environment (Di Baldassarre et al., 2011; Bates 2012; Schumann et al., 2009), but also provided a paradigm shift in flood modelling. In particular, there has been an increasing interest in evaluating potential of emerging remote sensing/earth observation data rather than keep on developing more sophisticated models (Bates 2012). Among the new data sources, airborne remote sensing products (e.g. Light Detection and Ranging (LiDAR) topography (Figure 1.2 and Figure 1.3), aerial photography, hyperspectral and Synthetic Aperture Radar (SAR) sensors), which provide high-resolution and high-accuracy data, are costly to acquire over large areas. For example, the vertical accuracy of LiDAR topography could normally reach 10-15 cm *RMSE* with spatial resolution of 2-5 m (Bates 2012). However, the cost of acquiring and processing the LiDAR data is around 750 US dollar per square kilometre (Humme et al., 2011). This is therefore a significant limitation when large-scale flood studies are undertaken, particularly in developing countries. The first major advances in PUB became possible thanks to considerable progress in orbital earth observing satellite technology. Those global low-cost earth observation (remote sensing) data provide large amount of scientific information, despite their often low-resolution, low-accuracy and low-frequency. There is a general consensus that the increased availability and quality of those low-cost remote sensing data will be valuable for improving prediction in ungauged basins (Hrachowitz et al., 2013), although the measurements provided by ground-based gauging stations still remain crucial for flood inundation modelling. Thus, integrating those new sources of low-cost data with ground-based observations and models as well as estimating associated uncertainties related to hydraulic modelling is a major scientific challenge.

*Figure 1.3 3-D LiDAR data shows downtown Manhattan (Image from NOAA)*

This thesis focuses on the potential and limitations of low-cost, space-borne data (e.g. freely available global DEMs, SAR imagery from ERS-1/2 and ENVISAT missions and radar altimetry) in supporting hydraulic modelling of floods in data-poor areas. Currently, most low-cost data are characterised by global coverage, easy accessibility, short repeat cycle at the cost of relatively coarse spatial resolution and low vertical accuracy (Schumann et al., 2009). However, as remote sensing technology is developing at fast pace, higher resolution and better accuracy data are expected to be globally available at low-cost in the near future (e.g. the recently launched Sentinel-1). Some research work that is reported in this thesis is also based on high resolution DEMs (e.g. LiDAR). These high resolution data are used to test the usefulness of low-cost data.

## 1.2   GLOBALLY FREELY AVAILABLE TOPOGRAPHY AS INPUT DATA FOR HYDRAULIC MODELLING

Topographic information is among the most important input data for hydraulic modelling (Farr et al., 2007). It is also considered as one of the most significant sources of uncertainty in hydraulic modelling (e.g. Jung et al., 2012a). There was a

lack of low-cost globally-available digital elevation model (DEM) before the launch of
the Shuttle Radar Topography Mission (SRTM, Figure 1.4). In February 2000, the
SRTM was successfully launched by a joint effort of the National Aeronautics and
Space Administration (NASA), the National Geospatial-Intelligence Agency (NGA)
and the German Aerospace Center (DLR). This 11-day mission flown on Space
Shuttle Endeavour offers a DEM of all land between 60 N and 56 S, about 80% of
Earth's land surface (Farr et al., 2007). The data processing of this interferometric
SAR technology product includes SAR focusing, motion compensation, coregistration
and interpolation, interferogram formation and filtering, and phase unwrapping
(Rabus et al., 2002). After the data processing, the SRTM data was sampled to the
resolution of 3 arc sec (approximately 90 m) globally and 1 arc sec (approximately 30
m) for the US territory (Farr et al., 2007), and recently also for Australia. The
primary goal of the mission was to produce a topographic dataset with globally
consistent and quantified errors (Rodriguez et al., 2006), with linear vertical absolute
height error of less than 16 m on average and linear vertical relative height error of
less than 10 m (Farr et al., 2007). The global validation of SRTM demonstrated that
the absolute height error in all continents ranged from 5.6 to 9.0 m at 90% confidence
(Table 1.1, Rodriguez et al., 2006). This validation mainly relied on the comparison
with kinematic GPS data.

***Figure 1.4*** *SRTM 90m Digital Elevation Data from CGIAR-CSI*
*(http://srtm.csi.cgiar.org/)*

### 1.2.1    Error characteristics of SRTM and its global assessment

The vertical absolute height error in Table 1.1 provides an overview of SRTM quality
at the continent level. This error is composed of several types of errors (Falorni et al.,
2005). Among those, the most characteristic error of the SAR-derived DEMs, like
SRTM, is random noise induced by radar speckles in the form of spikes and wells
(with a magnitude of approximately ±2~5 m, Hanssen, 2001). This random noise,
which affects the vertical elevation of SRTM data set, leads to relative height error

from 4.7 to 9.8 m at the continent level (Table 1.1). It is independent from one pixel to another and is linearly proportional to $1/\sqrt{n}$, where n is the number of pixels being aggregated (Rodriguez et al., 2006). Hydraulic modelling, in particular two-dimensional (2D) flood inundation modelling is hampered by the random noise on floodplain as the relative error dominates flood patterns and dynamics (Falorni et al., 2005). Some efforts on smoothing out the discontinuity and reducing this type error can be made either by using a wavelet filter (e.g. Falorni et al., 2005) or aggregating DEM cells (e.g. Neal et al., 2012a).

**Table 1.1** *Global assessment of SRTM DEM (Rodríguez et al., 2006)*

| Error Type | Eurasia | N. America | S. America | Africa | Australia | Islands |
|---|---|---|---|---|---|---|
| Absolute Height Error (m) | 6.2 | 9.0 | 6.2 | 5.6 | 6.0 | 8.0 |
| Relative Height Error (m) | 8.7 | 7.0 | 5.5 | 9.8 | 4.7 | 6.2 |

It is worth mentioning that DEM aggregation to resolutions much coarser than 90 m is only appropriate for very large-scale flood studies. Also, the SRTM elevations include vegetation canopy heights and removal of those is much less understood and only in a research phase meaning that the SRTM DEM is currently not available as a bare ground digital terrain model (DTM) by default. Due to the overestimated floodplain topography in heavy vegetated floodplains could lead to underestimation of inundation extent, vegetation needs to be removed before the DEM can be applied to flood inundation mapping or modelling (Baugh et al., 2013). Some studies focused on uniform vegetation height removal by inspecting the SRTM at independently known vegetation areas (Coe et al., 2008; Paiva et al., 2011). However, since the SRTM elevation represents a phase center height located between the bare ground and the top of the canopy (Brown et al., 2010), it would only be necessary to subtract a percentage of the total vegetation height. In particular, the percentage of canopy height that is removed from SRTM DEM should be quantified, as this percentage tends to differ from river to river. The new vegetation removal method also requires global vegetation height data (Lefsky, 2010; Simard et al., 2011) on SRTM. For example in Amazon, Baugh et al., (2013) showed subtracting 50 to 60 percentage vegetation height of the global vegetation dataset (Simard et al., 2011) from SRTM leads to optimum hydrodynamic modelling performance. These percentages in other major rivers of the world are not known at the moment thus more investigations are required.

Some scientific studies have highlighted that the vertical accuracy of SRTM is strongly influenced by terrain relief: large vertical errors and voids are frequent in high-relief terrain, while in the low- to medium-relief areas, the vertical errors are smaller and voids are much less (e.g. Sanders, 2007; Falorni et al., 2005). Floodplains, rivers and deltas, where hydraulic modelling usually takes place, are low relief areas and therefore associated with a lower absolute height error in SRTM data. For example, SRTM underestimates elevation 3.6-10.3 m at high-relief and high-altitude terrain for the Santa Clara River are in the US compared to national elevation data (NED, a patchwork of topographic data from several sources including LiDAR surveys). While SRTM differs at most 0.91 m for the pasture swath, 3.0 m for the riparian swath and 4.0 m for the commercial swath in the same area (Sanders, 2007). Falorni et al., (2005) showed the SRTM has a mean vertical error around 3.2 m when it was compared to Ground Control Points (GCPs) in a lowland and low relief catchment (the Little Washita Basin, Oklahoma, USA), while the vertical error magnitude was found to generally increase with increasing elevation (and slope) in the Tolt River Basin (Washington, USA). Patro et al., (2009) compared SRTM elevation values with topographic maps (toposheets from Survey of India (SOI)) in 153 spots of the Mahanadi River delta region (with elevation above sea level varying between 0 and 25 m). They found that the SRTM elevations are on average 2.3 m higher than those obtained from the topographic maps. Wang et al., (2012) compared the SRTM DEM elevations to gridded Global Positioning System (GPS) points in Longli Co Lake in southeastern Tibet of China and found large errors (Root Mean Square Error (RMSE) around 14 m). These vertical errors are higher than those of the two studies in Falorini et al., (2005) and Patro et al., (2009). This is likely attributed to the high relief and high elevation of the Tibetan Plateau.

One of the largest drawbacks of SRTM is that its SAR-based interferometer technology cannot obtain the geometry of the river bed below the water surface, but instead SRTM data provide water surface elevation in rivers at the time of the space shuttle overpass (February 2000; Farr et al., 2007). In addition, a coarse resolution pixel within SRTM also includes surrounding regions of the main channel. As a result, the channel bed elevation may be greatly overestimated in SRTM data (Figure 1.5).

The scientific community has proposed several approaches to deal with this issue (e.g. Neal et al., 2012a; Patro et al., 2009; Alfieri et al., 2014): reduce the SRTM bed elevation according to additional data sources; correct SRTM bed elevation using hydraulic geometry relationships, or assume a certain river flow at the time of shuttle overpass for SRTM data acquisition and adjust the inflow hydrograph accordingly. Alfieri et al., (2014) assumed the river bed elevation described by SRTM corresponds to average runoff conditions. Therefore, inflow hydrographs were reduced by subtracting the mean discharge, which was calculated from the 21-year discharge time series. Given the fact that the detailed topographic information on floodplain is rarely available in a large number of cases, as well as the overestimation of in-channel bed elevation of SRTM, the usefulness of the hybrid DEM (i.e. SRTM floodplain plus ground-surveyed in-channel geometry) in hydraulic modelling is also worth to explore. Patro et al., (2009) reduced the SRTM DEM-derived cross-section elevation values by 2.3 m, which was the average difference between the elevations of SRTM DEM and topographic maps, to simulate the river flows of the Mahanadi River in India with limited available data. Neal et al., (2012a) used hydraulic geometry theory to correct SRTM data by approximating river bed elevation in an 800-km reach of the River Niger in Mali. Hydraulic geometry theory relates river depth and width to river discharge using a series of power laws (Leopold and Maddock 1953). A derivation of the hydraulic geometry equation used within the subgrid channel model in Neal et al., (2012a) is given by:

$$d = \left( \frac{c}{a^{f/b}} \right) w^{f/b} \qquad\qquad\qquad (1.1)$$

where $d$ is the depth from the river bank elevation, $w$ is the bankfull width that now are available at two global database (Andreadis et al., 2013; Yamazaki et al., 2014), both the term $c/a^{f/b}$ and the exponent of w, $f/b$, are the free parameters that need to be estimated or calibrated (Neal et al., 2012a). Yan et al., (2015a) calibrated a SRTM-based hydraulic model on a reach of Danube River in which the bed elevation was treated as an additional parameter. The optimal bed elevation was the SRTM bed elevation values reduced by 5 m. Then the optimal model was used to predict another independent flood event, which yielded a Mean Absolute Error ($MAE$) of 1.37 m validated against water levels from four ground gauge stations; whilst this error is relatively large for commonly reported water level prediction accuracies, it is about the same as the actual integer precision of the SRTM DEM data.

The voids in the SRTM dataset which add more uncertainty in the terrain topology poses a considerable problem for hydraulic modelling. Recently, the voids of SRTM have been filled and two global coverage void-filled SRTM versions exist, from the Consortium for Spatial Information of the Consultative Group for International Agricultural Research (CGIAR-CSI, Jarvis et al., 2008) and the United States Geological Survey (USGS) Hydrological data and maps based on SHuttle Elevation Derivatives       at       multiple       Scales       (HydroSHEDS)       database (http://hydrosheds.cr.usgs.gov/). Recently in November 2013, the Land Processes Distributed Active Archive Center (LP DAAC) released the NASA SRTM Version 3.0 (SRTM   Plus)   product   collection   with   all   voids   eliminated   (available   at https://earthdata.nasa.gov/).

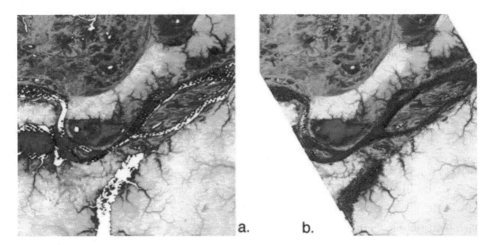

**Figure 1.5** *Images of water surface elevations from SRTM in the central Amazon Basin derived from the (a) C band and (b) X band systems. Compared to X band, C band elevations are missing (white areas) for some portions of the channel and lake areas. Elevation accuracies over water surfaces in both DEMs are degraded compared to surrounding land. Red represents the lowest elevations, followed by blue and yellow for the highest elevations.*

### 1.2.2    SRTM application in hydraulic modelling

Given the potential value of SRTM data for large-scale studies of alluvial rivers, floodplain and delta areas, a number of studies have explored the capability of SRTM in supporting large-scale hydraulic modelling (e.g. Sanders, 2007; Schumann et al., 2008, 2010; Alfieri et al., 2014; LeFavour and Alsdorf 2005; Neal et al., 2012a; Patro et al., 2009; Wang et al., 2012; Yan et al., 2013, 2014). These studies cover many aspects in hydraulic modelling including water level and water surface slope retrieval, flood extent simulation, water level and discharge prediction. The followed session summarizes those studies.

Wind roughening of the water surface and wave action can produce radar backscattering at large look angles (30 to 58 ) meaning that SRTM can yield elevation over water surfaces. This provides opportunity to estimate water surface slope from SRTM-derived water levels. Water slope and discharges calculated by LeFavour and Alsdorf (2005) on the Amazon River had only 6.2% difference compared to in-situ discharge measurement.

Apart from direct extracting water levels from DEM, a methodology for indirect water stage retrieval during floods also seems promising. This is to intersect high vertical accuracy DEM with airborne or space-borne imagery (e.g. Matgen et al., 2007; Oberstadler et al., 1997; Brakenridge et al., 1998) and derive water stages at the flood shoreline. This method was applied to global freely available DEM (e.g. SRTM) and timely low resolution SAR imagery as well. After intersecting SRTM DEM with Advanced Synthetic Aperture Radar (ASAR) Wide Swath Mode (WSM) imagery for 2008 flood in River Po of Italy, Schumann et al., (2010) compared the water level approximation with those derived from intersecting a high-resolution, high-accuracy LiDAR DEM with ASAR imagery. They were remarkably close to each other (with mean elevation error of the best fit line of -26 cm). A similar methodology was applied to the River Alzette in Luxembourg by Schumann et al., (2008), the SRTM derived water stages showed a relatively good performance with 1.07 m RMSE compared to those simulated by a flood inundation model calibrated with distributed ground-surveyed high water marks.

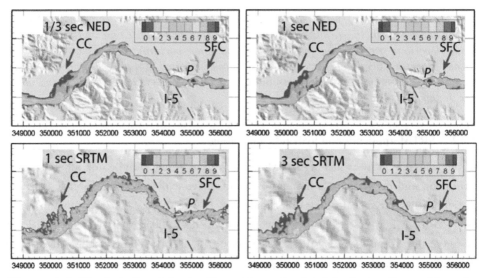

**Figure 1.6** *Flood depth of Santa Clara River simulated by hydraulic models using NED and SRTM datasets. Shown also are Castaic Creek (CC) and San Francisquito Creek (SFC) junctions, Interstate-5 (I-5) and point P which marks a constriction in the river (modify from Sanders 2007)*

Flood extent, as a proxy for flood damage, can be simulated by 2D hydraulic models. Those models are preferably built using high resolution, high accuracy topography and calibrated and/or validated on high quality observed data of flood extent (e.g. Hunter et al., 2005). Recent studies also tried to explore the potential of SRTM DEM on flood extent prediction (e.g. Sanders 2007; Yan et al., 2014). Relatively good performance could be obtained by simulating low-frequency floods if SRTM is appropriately treated. For example, Sanders (2007) found that SRTM topography yielded a 25% larger flood zone compared with the high-resolution topography (i.e. NED, national elevation data) in a steady-flow model application on the Santa Clara River (Figure 1.6). Wang et al., (2012) found that the HEC-RAS model built with SRTM yielded a 6.8% larger flood inundation extent and a mean water depth approximately 2 m shallower than that provided by a high precision topography.

**Table 1.2** *Global freely available (or low-cost) DEMs for hydraulic modelling (Schumann et al., 2014)*

| DEM | Spatial resolution | Vertical accuracy |
|---|---|---|
| SRTM | 90 m | 5.6-9 m |
| ASTER GDEM | 30 m | 7-14 m |
| ACE2 GDEM | 1 km | >10 m |
| GTOPO30 (HYDRO1k) | 1 km | 9-30 m |
| TanDEM-X | <12 m | <2 m |

The potential of SRTM DEM in simulating water levels and discharges in one-dimensional (1D) hydraulic modelling has also been explored. For example, Patro et al., (2009) showed a SRTM-based MIKE 11 model calibrated with in-situ river discharge and water level for the monsoon period in 2004 performed reasonable in model validation (with a water level RMSE less than 1 m). The model was validated using discharge and water level data for the same period in 2001 and 2002. The cross sections provided by SRTM were improved by referring to ground-surveyed topographic maps.

### 1.2.3    Other freely available or low-cost global DEMs

Apart from SRTM, there are also other free or low-cost global DEMs (Table 1.2). Advanced Space-borne Thermal Emission and Reflection Radiometer (ASTER) Global Digital Elevation Model (GDEM) is a 30-m spatial resolution DEM developed using stereo-photogrammetry. The global assessment of ASTER showed its accuracy is of 17 m at the 95% confidence level (Tachikawa et al., 2011), meaning that SRTM has some advantages over ASTER for most flood modelling studies (Bates et al., 2013). Although it has been recently improved in 2011 by reducing voids and correcting anomalies (http://www.jspacesystems.or.jp/ersdac/GDEM/E/4_1.html), very few studies have successfully used and tested this DEM product for inundation modelling and mapping (e.g. Gichamo et al., 2011; Wang et al., 2012). For example, Gichamo et al., (2011) found the ASTER GDEM is 27 m high for the stream central line of the River Tisza, Hungary. The vertical bias correction was therefore carried out by comparison of elevation points with a high accuracy DEM, produced a considerable improvement to the cross sections obtained. This bias correction was considered as a very important component in subsequent application; however, this cannot be done without a high accuracy DEM, which is rarely available in data-scarce areas. The ASTER DEM in River Po basin looks even worse (Figure 1.7): the elevation on the west part is systematically lower than that of the east part. The line divides two parts of the terrain showing systematic error can be clearly observed. Moreover, the in-channel part of River Po is abnormally higher than the floodplain outside the banks (see in Figure 1.3), which makes the end-user impossible to carry out any flood modelling based on ASTER DEM. This coincides the findings in Gichamo et al., (2011).

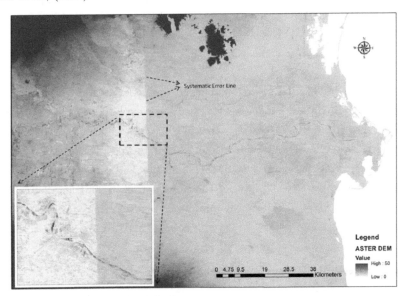

***Figure 1.7*** *ASTER DEM of River Po (Italy) Delta: ~8 m elevation difference is found on two sides of the 'systematic error line', in-channel elevation of River Po is ~20 m high than the surrounding floodplain.*

Other global DEMs such as Altimeter Corrected Elevations 2 (ACE2) GDEM and Global 30 Arc-Second Elevation (GTOPO30) have a resolution of 1 km, which is often too coarse for regional flood studies but may be adequate for large-scale applications when models can be run in sub-grid mode (Neal et al., 2012a; Schumann et al., 2013). TerraSAR-X add-on for Digital Elevation Measurement (TanDEM-X) at a spatial resolution of 12 m and a vertical accuracy of less than 2 m is certainly appealing, although the release of this version is currently under a commercial license. Also, scientific assessment of this DEM is at the moment only in an experimental phase and its use is thus fairly restricted, meaning that it is too immature to conclude on its use for hydraulic modelling. Having said that, the associated low-cost (100 euro per quota, ~7700 km$^2$) would be a relatively minor obstacle if cost were a constraint. First observations seems to indicate that TanDEM-X is a promising low-cost alternative and might allow for the first time more detailed local flood studies at the global scale (Yan et al., 2015b).

### 1.2.4    River width and depth database

As already stated earlier, global DEMs usually cannot provide accurate in-channel geometry information (e.g. main channel depth), which therefore needs to be approximated by using various methods (e.g. Neal et al., 2012a; Patro et al., 2009; Yan et al., 2015a). There is indeed a need to estimate river depths and widths with a consistent methodology for flood modelling based on freely available DEMs in data-scarce regions. To this end, global river bankfull width and depth databases were developed and are freely available (Andreadis et al., 2013; Yamazaki et al., 2014). The database by Andreadis et al., (2013) is derived as the following: the drainage areas for river reaches are derived from a 15 arc sec SRTM DEM globally (Lehner et al., 2008). The discharges were then estimated using the equation $Q = k\,A^c$, where $Q$ is discharge, $A$ is the drainage area, and $k$ and $c$ are coefficients (Wharton et al., 1989; Sweet and Geratz, 2003). Afterwards, regression equations were developed using the GRDC (Global Runoff Data Centre) data and discharge estimations, with the mean value of R2 equals to 0.95. The regression equations were used to estimate a mean annual peak flow (approximating the flow with 2 year return period) for every river reach delineated in the HydroSHEDS dataset. Then the widths and depths were calculated based on hydraulic geometry relationships, which describe the power law relation between discharges and widths (or depths). Even though errors in validation with widths derived from Landsat imagery ranged from 8 to 62%, this database can be useful to provide initial estimates for hydraulic or hydrologic modelling where other suitable measurements are not available. The global width database by Yamazaki et al., (2014) is calculated from satellite-based water masks and flow direction maps based on the SRTM Water Body Database and the HydroSHEDS flow direction map. The effective river width was compared with existing river width databases and relative differences were within ±20% for most river channels.

## 1.3   SATELLITE IMAGERY AND REMOTELY SENSED WATER LEVELS IN HYDRAULIC MODELLING

Hydraulic models produce various types of meaningful information for modellers and decision-makers, e.g. water level, flood extent, flood peak travel time etc. Among them, the flood extents provide the most intuitive and valuable information as a proxy for flood inundation hazard patterns in 2D form. In hydraulic modelling, the calibration is done to improve the fit between simulation and observation by adjusting model parameters, while the model performance can be evaluated through comparison to observation in model validation. The use of point data (zero-dimensional in space, such as water levels) to calibrate hydraulic models in simulating flood extent are fundamentally suspect (Horritt 2000), not to mention that floodplain-distributed water levels are hardly available in data-scarce areas. In the past, the lack of distributed flood extent information hampered the application of hydraulic models for flood extent prediction. Recent efforts on the integration of remotely sensed flood extent (e.g. SAR satellite imagery) as well as water levels (e.g. radar altimetry) with hydraulic models has revealed great potential of these data in supporting flood modelling (e.g. Smith 1997; Bates 2012, Schumann et al., 2009, Di Baldassarre et al., 2011). This section summarizes the current remote sensing data aiding flood model calibration and evaluation in data-sparse areas.

### 1.3.1   Low-cost satellite imagery for flood extent

*SAR imagery for flood monitoring*

There are multiple sources of remotely sensed flood extent information, such as aerial photography and thermal (optical) imagery. However, their disadvantages hamper their functionality for flood monitoring in data-scarce areas. For example, aerial photography is capable of providing high resolution imagery and is considered probably the most reliable source of remotely sensed flood extent data (e.g. Yu and Lane, 2006). However, large costs associated with airborne acquisitions make it less attractive in data-scarce areas. The visible and thermal bands of a flood image are obviously appealing for flood extent acquisition and there are some successful examples on flood mapping using such imagery (e.g. Marcus and Fonstad, 2008; the Moderate Resolution Imaging Spectroradiometer (MODIS)-based Dartmouth Flood Observatory record). However, these sensors cannot capture imagery under cloud cover which is commonly associated with flood events. Therefore, radar (i.e. microwave) remote sensing, particularly satellites carrying SAR sensors would be most useful for flood detection and monitoring since microwave signals can penetrate clouds and are reflected by the open water surface (e.g. Schumann et al., 2009), the insensitivity to weather conditions (e.g. Sanyal and Lu, 2004), as well as the ability to acquire data during day and night (e.g. Imhoff et al., 1986). Meanwhile, the costs of those images are fairly low (e.g. Di Baldassarre et al., 2011).

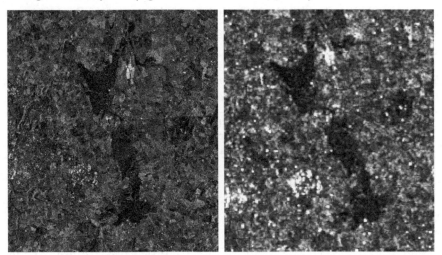

*Figure 1.8* *Satellite imagery capturing 2006 flood in River Dee, UK: ERS-2 SAR image (12.5 m resolution, left panel), ENVISAT ASAR image (75 m resolution, right panel) (Di Baldassarre et al., 2009)*

Two important aspects for satellite SAR imagery for flood modelling are image resolution and satellite revisit time. Various satellite missions provide imagery with various resolutions (mostly better than 100 m) and repeat cycle (from 11 to 46 days, see in Table 3) (Di Baldassarre et al., 2011; Schumann et al., 2009). Given there is an inverse relationship between image resolution and repeat cycle and the duration of flooding is commonly in the order of a few days, it is nearly impossible to capture more than one imagery per flood with the low repeat cycle satellites (e.g. TerraSAR-X, 11 days and RADARSAT-2, 24 days, Di Baldassarre et al., 2011).

However, this has been changed by satellite constellation missions, where several satellites working together to achieve reduced revisit time. For example, the recently launched Sentinel-1 mission (April, 2014) is designed as a two-satellite constellation, offering a global revisit time of just six days. The disaster monitoring constellation (DMC), initially launched in 2003, consists of a number of satellites, allowing daily imaging of any given point on the globe. The COSMO-SkyMed (COnstellation of small Satellites for Mediterranean basin Observation) system with its four satellite constellation can provide high resolution imagery with the revisit time as less as 2 hours (Covello et al., 2010). The emerging of those satellite constellation missions makes the acquisition of several images for a single flood event possible. For example, several flood imagery were acquired for flood events recently by COSMO-SkyMed in Italy (2009 event, 3 images), Albania (2010 event, 5 images), Pakistan (2010 event, 28 images, with 11 processed), Thailand (2010 event, 2 images) and Australia (2011 event, 2 images) (Pierdicca et al., 2013). It has also been successfully used in various applications in the field of risk and emergency management, such as Myanmar and Haiti flood.

**Table 1.3** *Current and past satellite missions that could be used for flood extent monitoring (Di Baldassarre et al., 2011; Schumann et al., 2014)*

| Mission/Satellite | Launch year | Spatial resolution | Repeat cycle (days) | Cost |
|---|---|---|---|---|
| ALOS | 2006 (finished in 2011) | 10-100 m | 46 | Low: €600 for three resolutions (10 m, 30 m, 100 m, Archive) |
| ENVISAT ASAR | 2002 (finished in 2012) | 12.5-1000 m | 35 | Free/Low: From €150 of 150 m resolution to €500 of 25 m resolution |
| TerraSAR-X* | 2007 | 1-40 m | 11 | High: From €1750 of 40 m resolution, to €5950 with 1 m resolution (New acquisition) |
| RADARSAT-2* | 2007 | 1-100 m | 24 | High: From $3600 of 100 m resolution, to $8400 of 1 m resolution (New acquisition) |
| COSMO-SkyMed* | 2007 | 1-100 m | 16 (2 hours with four-satellite constellation) | High: From €1650 of 100 m resolution, to €9450 of 1 m resolution (New acquisition) |
| Sentinel-1 | 2014 | 5-100 m | 12 (6 with two-satellite constellation) | Free/Low |
| ERS-1 | 1991 (finished in 2000) | 25 m/150 m | 35 | Low: €90 for 150 m resolution, €180 for 25 m resolution (Archive) |
| ERS-2 | 1995 (finished in 2000) | 25 m/250 m | 35 | Low: €90 for 150 m resolution, €180 for 25 m resolution (Archive) |

Fast acquisition and delivery maps from those satellite constellations can also be produced in a fully automatic way. On the other hand, several satellite missions designed for high resolution image acquisition also provide the service of obtaining low resolution imageries with low cost and high repeat cycle (Table 1.3, Figure 1.8). Low resolution imagery are invaluable in studies where multiple images are required. For example, Environmental Satellite (ENVISAT) ASAR WSM revisit time could be of the order of 3 days and can be quickly obtained (~24 h) at no (or low) cost (Di Baldassarre, et al., 2009a). The resolution of those SAR imagery seems to satisfy most large-scale flood modelling studies in data-scarce areas, where global DEMs are used for hydraulic model building. Currently the only reliable global DEM (i.e. SRTM) has a resolution of ~90 m, which is similar to that used for a model building and calibration.

*New techniques for flood detection from SAR*

Before integrating SAR imagery and hydraulic models, SAR images have to be processed to retrieve meaningful flood boundaries. There are many SAR imagery-processing techniques to delineate flood extent, which have been reviewed by several authors (e.g. Schumann et al., 2009; Di Baldassarre et al., 2011; Liu et al., 2004, Lu et al., 2004). Various methods include visual interpretation (e.g. MacIntosh and Profeti, 1995), image histogram thresholding (e.g. Brivio et al., 2002), image texture algorithms (e.g. Schumann et al., 2005), automatic classification algorithms (e.g. Hess et al., 1995), and image statistics-based active contour models (e.g. Horritt, 1999, Figure 1.9). Each method has its own pros and cons. Users could choose the appropriate method according to the characteristic of the flood event and study area.

Recently, a new method for flood extent delineation has been developed. Automatic thresholding algorithms has been improved since the launch of the high resolution SAR satellites such as TerraSAR-X, Radarsat-2 and the COSMO-SkyMed (e.g. Mason et al., 2012, Giustarini et al., 2012, Martinis et al., 2009; Pulvirenti et al., 2012; Matgen et al., 2011; Schumann et al., 2010). In particular, Pulvirenti et al., (2011) developed a new algorithm to detect not only open water surfaces, but also forested and agricultural flooded areas from COSMO-SkyMed SAR imagery based on a fuzzy classification approach. This algorithm takes into account three well-established electromagnetic models of surface scattering, with simple hydraulic considerations and contextual information. For flood extent delineation of high resolution SAR imagery in urban areas, Matgen et al., (2011) and Giustarini et al., (2013) presented a method relying on the calibration of a statistical distribution of "open water" backscatter values inferred from SAR images of floods. They combined radiometric thresholding, region growing, and change detection as an approach enabling the automatic and reliable flood extent extraction. For urban flooding issues,

Mason et al., (2014) investigated whether urban flooding can be detected in layover regions where flooding might not be apparent, using double scattering between the possibly flooded ground surface and the walls of adjacent buildings. The method could lead to improved detection of flooding in urban areas depending on the particular flooding situation.

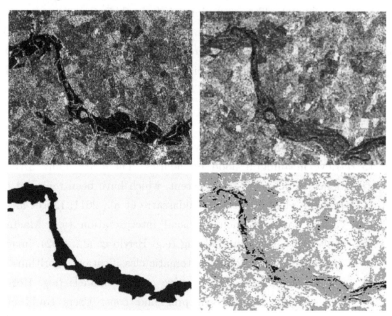

***Figure 1.9*** *SAR imagery and flood extent delineation results. The RADARSAT imagery (top left) is segmented using the active contour model ('snake') to yield the inundation extent. The ERS-2 imagery (top right) is threshold into wet/dry/undetermined classes (bottom right). RADARSAT imagery and ERS-2 imagery are provided by Canadian Space Agency and European Space Agency, respectively. (Figure extracted from Horritt 2006)*

Thanks to the development of the flood extent delineation techniques, the flood monitoring systems based on those algorithms are established on the recent satellite missions. For example, Martinis et al., (2013) presented a combined two-phase flood monitoring system (i.e. MODIS and TerraSAR-X) to support flood disaster management. This system provides flood imagery at different spatial resolutions and time-critical on demand acquisition, as well as the pre-processing and web-based platform for imagery dissemination in near-real time. Westerhoff et al., (2013) presented an automated surface water detection technique from SAR, with the output maps uploaded on an open data server. The algorithm is to match the backscatter distribution to the ones of training data set of dry land and surface water. This systematic and automated processing algorithm could match well with Sentinel-1 satellite mission, which would together form the so-called Global Flood Observatory, provide SAR-based flood extent imagery globally in near-real time. Martinis et al., (2014) presented a fully automated processing chain for near real-time flood detection using high resolution TerraSAR-X SAR data. The processing chain including SAR data pre-processing, computation and adaption of global auxiliary data, unsupervised initialization of the classification as well as post-classification refinement by using a fuzzy logic-based approach. The dissemination of flood maps resulting from this service is performed through an online service which can be activated on-demand for emergency response purposes. The flood maps delivered by the above-mentioned flood monitoring systems can also be used for flood model calibration, as well as the validation of flood forecasting systems.

*SAR-derived flood extent for model calibration and evaluation*

Many studies have successfully utilized satellite imagery in flood extent prediction using various performance measures (e.g. Aronica et al., 2002; Di Baldassarre et al., 2009a and 2010; Pappenburger et al., 2007; Horritt and Bates., 2001; Bates et al., 2004; Brandimarte et al., 2009; Prestininzi et al., 2010, Table 1.4). Particularly in data-scarce areas, moderate and low resolution SAR can achieve model calibration by identifying the optimal (distributed) roughness coefficients for 1D hydraulic simulations (e.g. Di Baldassarre et al., 2009a), 2D simulations (e.g. Tarpanelli et al., 2013; Hostache et al., 2009; Mason et al., 2009) and sophisticated finite element/volume models (e.g. Horritt, 2000; Horritt et al., 2007). In particular, ENVISAT ASAR coarse resolution imagery can verify the model against a past flood event in near real time after its acquisition when a simple 1D hydraulic model was used (Di Baldassarre et al., 2009a). Given the model structure is never perfect in hydraulic modelling, model users should choose the appropriate hydraulic codes according to the study needs. ENVISAT ASAR imagery was also used to assess hydraulic model performance, support hydraulic structure and code selection (e.g. Prestininzi et al., 2010). Furthermore, SAR imagery also contributes to produce fuzzy flood risk map (e.g. Schumann and Di Baldassarre, 2010; Merz et al., 2007), which can lead to a sustainable and affordable flood risk mitigation plan (Brandimarte et al., 2009). To assess various sources of uncertainties which are intrinsic to any hydraulic modelling exercise, low-cost SAR image also play an important role (e.g. Hunter et al., 2005;). Di Baldassarre et al., (2010) demonstrated the advantage of probabilistic flood mapping over deterministic approaches. In this study, European Remote Sensing-2 (ERS-2) SAR imagery (with ground resolution of 25 m) was used as observation data to assign weights to the model ensembles in a Generalized Likelihood Uncertainty Estimation (GLUE, Beven and Binley, 1992) framework.

Afterwards, probabilistic flood maps which account for uncertainties in inundation modelling were produced based on those ensembles. In addition, the poor-resolution flood extent imagery (ERS-2 SAR and RADARSAT with 25 m ground resolution) was found to be useful in constraining hydraulic model calibration in a Bayesian updating procedure of simulating two events in the upper River Severn, UK (Bates et al., 2004). In fact, explicitly assess uncertainties associated with both model and SAR imagery can be achieved by either a reliability diagram which was proposed by Horritt 2006 (e.g. Di Baldassarre et al, 2009b), or by using a fuzzy set approach (Pappenberger et al., 2007; Matgen et al., 2004). Overall, low-cost SAR satellite imagery can contribute to flood inundation modelling in various aspects such as identification of optimum model parameters and better model structures, development of flood risk mitigation plan and flood risk maps, assessment of uncertainties in flood inundation models as well as flood model verification in near real time.

In addition, there are also examples of integrating water elevation changes derived from the Interferometric Synthetic Aperture Radar (IfSAR) imagery in hydraulic model calibration (e.g. Jung et al., 2012b; Alsdorf et al., 2005). The temporal (dh/dt) and spatial (dh/dx, dh/dy) variations of water levels derived from the Advanced Land Observing Satellite (ALOS) Phased Array L-band Synthetic Aperture Radar (PALSAR) interferometry selected only slightly different values for channel Manning`s n for a 2D hydrodynamic model calibration in a major tributary of the Mississippi (Jung et al., 2012b). This new calibration strategy indicated the potential of using IfSAR imagery for enhanced prediction and assessment of future flood events (Figure 1.10). In a simpler case, Alsdorf et al., (2005) used the dh/dt of water level values derived from IfSAR to simulate floodplain storage in a linear diffusion hydraulic model of the Amazon. The model captured the fundamental behavior of the hydrograph recession of all three reaches studied.

*Table 1.4* *Applications of space-borne SAR image in hydraulic modelling in recent years*

| SAR image | Ground resolution | Modelling purpose and findings | Case study | Reference |
|---|---|---|---|---|
| ENVISAT ASAR | 150 m | Calibrate hydraulic model for discharge and water level simulation | Tiver River, Italy | Tarpanelli et al., (2013) |
| ENVISAT ASAR | 150 m | Select the most appropriate hydraulic model structure when the simuated flood events are with different magnitude | River Po, Italy | Prestininzi et al., (2010) |
| Terra ASTER | 90 m | Propose and formulate a sustainable and affordable flood risk mitigation plan | City of Jimani, Dominican Republic | Brandimarte et al., (2009) |
| ERS-1 SAR | 25 m | Identify the optimum distributed roughness coefficients in a finite element model calibration | River Thames, UK | Horritt, (2000) |
| ENVISAT ASAR | ~100 m | Verify the 1D hydraulic model in near real time after its acquisition | River Po, Italy | Di Baldassarre et al., (2009a) |
| ERS-2 SAR | 25 m | Demonstrate the advantage of probabilistic flood mapping over deterministic approaches by accounting uncertainties in hydraulic modelling | River Dee, UK | Di Baldassarre et al., (2010) |
| ERS-2 SAR and RADARSAT | 25 m | Constrain hydraulic model calibrations in a Bayesian updating procedure and discover the broadly consistent optimum friction values of two events | River Severn, UK | Bates et al., (2004) |
| ERS-2 and ENVISAT | 25 m and 150 m | Produce an event-specific fuzzy flood risk map by fusing the SAR imageries with vulnerability-weighted land cover map | River Dee, UK | Schumann and Di Baldassarre (2010) |
| ERS-1 SAR | 25 m | Constrain uncertainties in hydraulic models in a GLUE framework | River Thames, UK | Aronica et al., (2002) |
| ENVISAT ASAR | 12.5 m | Take account of the uncertainty in the SAR image using a fuzzy set approach | River Alzette, Luxembourg | Pappenberger et al., (2007) |
| ERS-2 SAR and ENVISAT ASAR | 25 m and 150 m | Explicitly estimate observed flood extent uncertainty in inundation modelling using reliability diagram | River Dee, UK | Di Baldassarre et al, (2009b) |

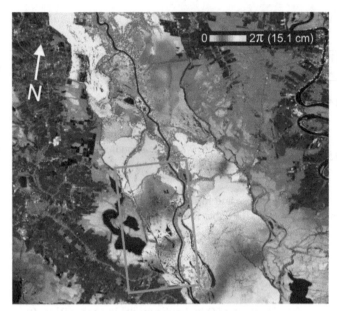

**Figure 1.10** *Differential wrapped interferogram of L-band PALSAR superimposed on floodplain of the Atchafalaya River, USA. The orange rectangular shows the flood modelling domain. The colour scale represents one cycle of interferometric phase (interpreted as 15.1 cm in vertical displacement). These fringes represent water level changes between 16 April 2008 and 1 June 2008 (Image extracted from Jung et al., 2012b).*

### 1.3.2   Freely available radar altimetry for water level

Recent developments of remote sensing techniques have demonstrated that water surface elevation, water surface slope, and temporal change can be measured from space (Alsdorf et al., 2007). Radar altimeter now has routinely measured the surface of fresh water bodies albeit its original purpose was for measuring ocean surface elevations (e.g. Koblinsky et al., 1993; Birkett 1998; Getirana et al., 2009; Santos da Silva et al., 2010). The altimetric water level measurements can be considered as a space-borne virtual gauge which provides discrete measurements at the satellite ground track (Birkinshaw et al., 2010). Due to the overshoot problems when radar signals target from land to water surface, radar altimeters are more accurate at measuring water levels of wider rivers. Vertical elevation accuracy of radar altimetry over river surfaces are ~10 cm at best and are usually ~50 cm, while the accuracies improve to 3-4 cm Root Mean Square (*RMS*) with the increased averaging over areas larger than 100 km$^2$ (e.g. Birkett et al., 1995, 2002; Maheu et al., 2003; Hwang et al., 2005; Frappart et al., 2006; Kouraev et al., 2004, Table 1.5). However, in some cases the discrepancies between altimetry and in-situ level measurements might be high and can be in the order of 2 m (Birkinshaw et al., 2010). This error depends on the type of sensor used and the distance between the sensor and virtual station, where radar satellite tracks intersect with a river reach and the ground-based gauging station. Due to reduced pulse averaging and differing echo shapes (Alsdorf et al., 2007), their relatively large orbital spacing (in both time and space) are a major limitation for the integration of this type of information for hydraulic modelling.

**Table 1.5** *Currently available radar altimetry for inland water level monitoring*

| Mission/Satellite | Launched Time | Vertical Accuracy | Repeat Cycle (days) | Reference |
|---|---|---|---|---|
| ERS-1 | 1991 | not applicable | 35 | Santos da Silva et al., 2010 |
| ERS-2 | 1995 | > 0.20 m | 35 | Domeneghetti et al., 2014; Santos da Silva et al., 2010 |
| ENVISAT | 2002-2008 | > 0.15 m | 35 | Frappart et al., 2006; Yan et al., 2015a; Domeneghetti et al., 2014; Santos da Silva et al., 2010 |
| TOPEX/Poseidon | 1992-2005 | > 0.30 m | 10 | Birkett et al., (1995), (2002); Maheu et al., 2003; Kouraev et al., (2004); Hwang et al., 2005 |
| Jason-1 | 2001 | no river data | 10 | Santos da Silva et al., 2010 |
| Jason-2 | 2008 | no validation published | 10 | Calmant et al., (2009) |
| Cryosat-2 | 2010 | no validation published | 369 days with 30 day sub-cycle | Moore et al., (2014) |

Some studies have focused on the integration of radar altimetry and hydraulic models (e.g. Hall et al., 2010; de Paiva et al., 2013; Yan et al., 2015a; Biancamaria et al., 2009; Domeneghetti et al., 2014). The high accuracy of altimetry data provided by the latest satellite missions and the promising results obtained in recent applications suggest that these data may be employed in the calibration and evaluation of hydraulic models. For instance, radar altimetry was applied in a large-scale 2D hydrodynamic model validation study for the first time by Wilson et al., (2007). The comparison of model simulations with radar altimetry yielded a *RMSE* of 0.99 m for the water stage of high flow over a 22-month period of the Amazon River. However, the RMSE increased to 3.17 m for water stages at low flow. More recently, another hydrological/hydrodynamic modelling study of the Amazon River showed the model performed well at most ENVISAT-derived altimetry virtual stations (60%) in terms of water level prediction. Larger biases were found for some stations ranging from -3 to -15 m. This is likely due to the SRTM DEM used to condition the model (de Paiva et al., 2013). Similarly, a large-scale coupled hydrologic and hydraulic modelling technique was used by Biancamaria et al., (2009) to model the Ob River in Siberia. Comparison of modelled water level with TOPEX/Poseidon (T/P) altimetry data allowed the estimation of parameters in the hydraulic model (i.e. river depth and Manning`s coefficient). In another study, the rating curves for 21 virtual stations in the upper Negro River were constructed by using water level derived from T/P and ENVISAT radar altimetry and discharge simulated by the Muskingum-Cunge approach (Leon et al., 2006). The estimated water depth for zero effective flow using the rating curves were found to be reliable compared to those measured. This work highlighted a promising application of orbital altimetry for large river where hydrometric data are not always available.

## 1.4   UNCERTAINTIES AND PROBABILISTIC FLOOD MAPPING

In recent years, there has been increasing interests in assessing uncertainty in hydrology and flood modelling. Understanding and quantifying those uncertainties is a key issue in uncertainty communication with decision makers (Montanari and Brath, 2004). Many studies have described that there are several sources of uncertainties intrinsic to flood inundation modelling, such as model structure, topographic data, model parameter and inflow etc (e.g. Aronica et al., 2002; Pappenberger et al., 2006; Di Baldassarre and Montanari, 2009). Among those, roughness is usually used as calibration parameter in flood inundation modelling given a predetermined model structure, topographic data and simulated flood event with inflow data. Due to the fact that all of those components in the modelling system are unavoidably associated with uncertainties, parameter uncertainty tends to become a proxy of them. To some extent, parameters are used to compensate other sources of uncertainties during the calibration process. Ultimately, in the author`s opinion, the understanding of parameter uncertainty could be a start to understand the whole modelling system. On the other hand, given the understanding of parameter uncertainty should be provided by a model with fixed model structure, inflow, topography etc, a methodology that looks at several major sources of uncertainty should be considered.'

Among the major sources of uncertainty mentioned above, the only one directly provided by remote sensing data is the topography (i.e. SRTM or other global DEM). As the objective of this thesis is to evaluate the potential of those data in hydraulic modelling of floods, the topographic uncertainty is considered by benchmarking with the high resolution, high accuracy DEM. In other words, SRTM and its uncertainty are not examined using any (explicit) sampling-based approach, but rather (implicit) benchmarking approach. Meanwhile, other sources of uncertainty are explicitly examined by a Monte Carlo-based approach.

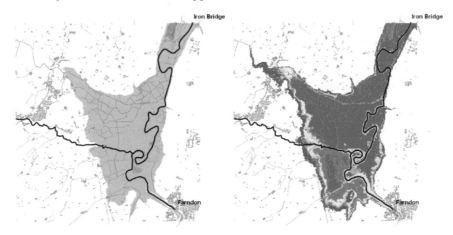

*Figure 1.11* *1-in-100 year flood inundation map: deterministic map (left panel), probabilistic map taking into account uncertainty (right panel, probability of inundation decreases from red, 1, to white, 0). This figure is extracted from Di Baldassarre et al., (2010)*

The flood inundation maps are often produced by hydraulic models using deterministic or probabilistic approach. The deterministic flood maps which are produced by using a fully 2D physically-based best-fit model, are precise, but potentially wrong, due to the fact that they ignore the above mentioned uncertainties in inundation modelling. There is perhaps a philosophical question here: 'choose to be approximately right, or precisely wrong'. The deterministic map which demonstrates the exact dry and wet land would make the decision making more straightforward, but not necessarily correct. On the contrary, the probabilistic map would leave more space to the decision making process, as it indicates the probability of inundation during floods. The probabilistic flood maps which explicitly consider various sources of uncertainties are believed to be theoretically more appropriate for visualizing flood hazard (Figure 1.11, Di Baldassarre et al., 2010), even though the interpretation of this 'probability' and uncertainty communication to the decision makers are not easy, as well as their application in flood risk studies are still limited. Therefore, the handling of uncertainty in flood modelling should be more explicit and transparent, perhaps in an identified framework that is stationary for various flood events.

## 1.5   OBJECTIVES

The main objective of this thesis is to explore the potential of low-cost space-borne topographic, flood extent and water level data in supporting hydraulic modelling. Some of those data (e.g. SAR imagery) has been examined in the literature rather in an isolated manner, therefore, in this thesis we investigate their utility for hydraulic modelling using a holistic view. The specific objectives are as following:

- To explore the potential of global freely available DEM (i.e. Shuttle Radar Topography Mission, SRTM) to support hydraulic modelling of floods.

- To explore the potential of low-cost space-borne satellite imagery and radar altimetry for hydraulic model calibration and evaluation.

- To investigate to what extent the topographic errors affect hydraulic modelling in view of other major sources of uncertainty that are intrinsic to any modelling exercise.

- To investigate how topographic data with different resolutions affect hydraulic modelling

- To investigate the appropriateness of regional and physically-based approaches for estimating design floods to map flood hazard in data-scarce areas.

## 1.6   METHODOLOGY

### 1.6.1   Research approach

To explore the usefulness of low-cost, space-borne data in supporting hydraulic modelling of floods, the research work has been carried out using a model calibration-evaluation approach. Moreover, topographic uncertainty is assessed in view of many other major sources of uncertainties (Chapter 2 and 3). This methodological approach is applied to different case studies. Thus, specific modelling exercises are implemented with slight, tailor-made modifications to deal with practical issues, such as the actual data availability, the characteristics of flood events. The methodological steps followed in this thesis are summarized in Figure 1.12 and Table 1.6.

Low-cost, space borne topographic and flood extent data are evaluated in the two river reaches of the Dee (Chapter 2) and the Po (Chapter 3), while topographic and water level data are considered in Danube River (Chapter 4). In the three cases of River Dee (Chapter 2), River Po (Chapter 3) and Danube (Chapter 4), hydraulic models are built to simulate particular flood events based on SRTM topography and their inflow hydrographs. For the cases with two independent events available (Chapter 3 and 4), we follow a rigorous model calibration and evaluation approach. Otherwise only calibration event is considered (Chapter 2). Uncertainty analysis is carried out to take account model uncertainties (Chapter 2 and 3). For the cases with LiDAR dataset available (Chapter 2 and 3), SRTM-based model results are benchmarked with that derived from LiDAR-based model. Otherwise, the performance of SRTM-based model is assessed by comparing to in-situ data (Chapter 4).

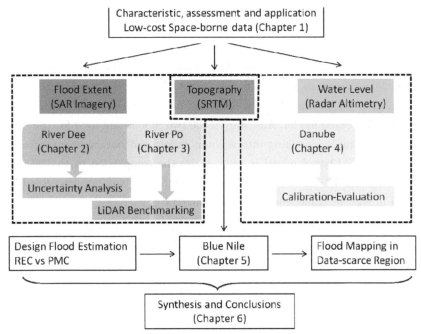

***Figure 1.12*** *An overview of this thesis work*

In the River Dee case (Chapter 2), inundation model is built based on SRTM topography and in-situ inflow data during the flood event (i.e. year 2006). The usefulness of SRTM-based model is evaluated through benchmarking with a high-resolution, high-accuracy LiDAR-based model. Given the fact that parameter uncertainty is the major source of uncertainty in this study, the comparison between two models is made in view of parameter uncertainty using Monte Carlo framework. The two models are conditioned by the satellite image indicating the flood extent of the year 2006 event.

**Table 1.6** *Overview of case studies of this thesis: Data investigated, flood events and uncertainty considered, and scale of the study*

| Case Study | Data investigated | | | Flood event considered | | Uncertainty Analysis | Scale |
|---|---|---|---|---|---|---|---|
| | SRTM | SAR Imagery | Radar Altimetry | Calibration event | Evaluation event | | |
| River Dee (Chapter 2) | X | X | | X | | X | Medium |
| River Po (Chapter 3) | X | X | | X | X | X | Medium ~ Large |
| Danube River (Chapter 4) | X | | X | X | X | | Large |

In the River Po case (Chapter 3), hydraulic models are built based on SRTM topography for two flood event: year 2000 flood event as the calibration event while year 2008 flood event as the evaluation event. During the model calibration and evaluation, the SRTM-based model is benchmarked with a high-resolution, high-accuracy LiDAR-based model. Afterwards, the two models are used for the estimation of design flood profile by taking into account inflow and parameter uncertainty. The usefulness of SRTM-based model is assessed by the comparison to the LiDAR-based model during model calibration, model evaluation, and design flood profiles estimation in view of inflow and parameter uncertainty.

In the Danube case (Chapter 4), as there is a lack of LiDAR topographic data, hydraulic models are built based on SRTM topography only. The SRTM-based model is calibrated by using radar altimetry water levels in the 2006 event. Afterwards, the calibrated model with optimum parameter set is evaluated by simulating the 2007 event, where the water levels used for evaluation are measured in the ground gauge stations. The inflow information for both 2006 and 2007 event also comes from ground gauge stations.

The Blue Nile case (Chapter 5) is an application of inundation modelling based on SRTM topography, which focuses on design flood estimation by using a traditional approach (i.e. regionalization) and a more relatively new approach (i.e. physically-based model chain). In particular, design flood inundation estimated by using regional envelope curve (REC) and physical model cascade (PMC) are benchmarked with that estimated from in-situ data. Potential and limitations of those two methods for design flood estimation in data-scarce areas are discussed.

### 1.6.2   Modelling tools

Over the past decades, the scientific literature has presented a plethora of hydraulic codes to simulate flood propagation and inundation processes. On the one hand, simple, one-dimensional (1D) models (e.g. kinematic wave) are often seen as not being suitable for modelling inundation processes as flow directions in the floodplain areas are much more complex. On the other hand, complex, fully three-dimensional (3D) models suffer of numerical stability, computational time and availability of input data (Dottori et al., 2013). Hunter et al., (2007) pointed out that one should carefully choose the modelling tools by considering the model structure and assumptions, data requirement, modelling purpose, computational time etc. In this thesis, 1D model (i.e. HEC-RAS) is only used in Po River case where flow conditions of simulated event fulfil the assumption of the 1D St. Venant equations, while HEC-RAS results also suit the type of available data in model calibration. It is worth to mention that HEC-RAS is not suitable for simulating a 2D process-dominated event due to model structural uncertainty (show in Chapter 3). Using HEC-RAS in such a situation often leads to over-estimation of maximum flood width. Having said that, flood modelling based on low-cost data using simple 1D model such as HEC-RAS is also of great interest to scientists and practitioners. Thus, this thesis follows a compromise between the needs for both physical realism and parsimony and explores the potential and limitations of remote sensing data by using the following two model codes: LISFLOOD-FP and HEC-RAS.

*LISFLOOD-FP*

Between the simple 1D and complex 3D models mentioned above, LISFLOOD-FP solves a simplified version of the shallow water equations that preserve acceleration but neglect the advection term, as advection is relatively unimportant in many floodplains flows (Hunter et al., 2007). The model comprises two equations that calculate the continuity of mass in each cell (Equation 1.2) and continuity of momentum between cells (Equation 1.3) in the direction of x. Although arranged in two dimensions, the momentum equation implemented across each face of each grid cell is in fact a one-dimensional calculation such that the fluxes through each cell face are decoupled from each other. It provides the water depth and discharge for each time step in each cell based on the raster grid being used. The equations are solved using an explicit forward finite difference scheme (e.g. Neal et al., 2012b; Hunter et al., 2007).

- Continuity

$$\frac{\partial Q}{\partial x} + \frac{\partial A}{\partial t} + q = 0 \qquad (1.2)$$

- Momentum

$$\frac{\partial Q}{\partial t} + \frac{gA\partial(h+z)}{\partial x} + \frac{gn^2 Q^2}{R^{4/3} A} = 0 \qquad (1.3)$$

Where $Q$ $[L^3T^{-1}]$ is the discharge, $A$ is the flow cross section area $[L^2]$, $x$ is the distance between cross sections $[L]$, $t$ is the time $[T]$, $q$ is a lateral inflow term $[L^3T^{-1}]$, $z$ is the bed elevation $[L]$, $R$ is the hydraulic radius $[L]$, $g$ is the acceleration of gravity $[LT^{-2}]$.

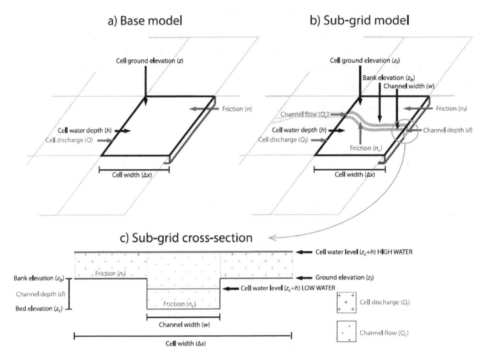

**Figure 1.13** *Conceptual diagram of (a) LISFLOOD-FP base model, (b) subgrid channels model, and (c) subgrid cross section (Figure extracted from Neal et al., 2012).*

To better integrate coarse resolution global DEMs with LISFLOOD-FP, Neal et al., (2012) developed the subgrid approach, which allows the simulation of the scenarios that main river channel width is smaller than the grid size (Figure 1.13). The subgrid approach allows the modeller to specify channel width, channel depth as well as the bank elevation inside each cell so that it can better emulate flood propagation in the main channel for coarse resolution models. It was proved to change the floodplain inundation dynamics significantly and increase simulation accuracy in terms of water levels, wave propagation speed and inundation extent compare with the pure 1D channel model or 2D floodplain model of LISFLOOD-FP (Neal et al., 2012). The subgrid approach uses the floodplain flow model of Bates et al., (2010), which introduced the local inertial term to the diffusive wave equation to significantly reduce the computation cost.

*HEC-RAS*

Hydrologic Engineering Centers River Analysis System (HEC-RAS) is a 1D hydraulic model that can simulate steady and unsteady flows in river channels including floodplains. HEC-RAS has been widely used for hydraulic modelling (e.g. Pappenberger et al., 2005, 2006; Schumann et al., 2007; Brandimarte et al., 2009), and a number of studies have proven its reliability in simulating floods in natural rivers (e.g. Horritt & Bates 2002; Castellarin et al., 2009) compared with 2D models (LISFLOOD-FP, TELEMAC-2D). However, it should be noted that HEC-RAS cannot reproduce the interaction process between the main channel and floodplain, the 2D process on the floodplain, as well as minor hydraulic effects such as secondary circulations and high-order turbulence. Often, these minor effects can be neglected considering the accuracy requirements in large-scale problems and modelling purposes.

HEC-RAS uses Equation (1.2) to preserve the continuity of mass. Unlike LISFLOOD-FP, the conservation of momentum in HEC-RAS is presented by momentum equation in full St. Venant equations (Equation 1.4):

$$\frac{\partial Q}{\partial t} + \frac{\partial}{\partial x}\left[\frac{Q^2}{A}\right] + \frac{gA\,\partial(h+z)}{\partial x} + \frac{gn^2 Q^2}{R^{4/3} A} = 0 \qquad (1.4)$$

Where $Q$ [L³T⁻¹] is the discharge, $A$ is the flow cross section area [L²], $x$ is the distance between cross sections [L], $t$ is the time [T], $z$ is the bed elevation [L], $R$ is the hydraulic radius [L], $g$ is the acceleration of gravity [LT⁻²].

### 1.6.3    Model performance measures

In this thesis, the two types of information are obtained from the modelling results, which are also usually provided by hydraulic models in general: (1) spatially distributed flood extent, (2) at-a-point time series of flood water stages or flood profiles along the river reach. Performance measures need to be selected to quantitatively assess model performances.

In this thesis the water level hydrographs and flood profiles are evaluated by using Root Mean Square Error (*RMSE*) or Mean Absolute Error (*MAE*). For evaluating flood extent, many performance measures are described in the literature for various circumstances (e.g. Aronica et al., 2002; Schumann et al., 2009). Among those, binary pattern measure (*F*) is a relatively unbiased measure that simply and equitably discriminates between under-prediction and over-prediction. As such, the optimal simulations will provide the best compromise between these two undesirable situations. Therefore, we use *F* to evaluate the simulated flood extent in this thesis:

$$F = \frac{A}{A + B + C} \tag{1.5}$$

where $A$ is the number of cells correctly simulated by the model, $B$ is the number of cells simulated as wet that is observed dry (i.e. over-prediction), $C$ is the number of cells simulated as dry that is observed wet (i.e. under-prediction). $F$ ranges from 0 to 1, the higher the better (Aronica et al., 2002; Horritt et al., 2007, Stephens et al., 2013).

## 1.7   OUTLINE OF THIS THESIS

To explore the potential and limitations of low-cost, space-borne data for flood inundation modelling, this thesis is structured in six chapters.

Chapter 2 presents a case of exploring the usefulness of SRTM topography and ERS-2 image under parameter uncertainty in a medium river: River Dee in UK.

Chapter 3 explore the value of SRTM topography and ENVISAT satellite image in a medium-to-large river: River Po in Italy. The design flood profiles produced by SRTM-based model are benchmarked with LiDAR-based model under inflow and parameter uncertainty.

Chapter 4 investigate the usefulness of SRTM topography and ENVISAT radar altimetry in supporting inundation modelling in a large river: Danube in Europe. A model built and calibrated by using globally freely available data is evaluated by in-situ water level measurements.

Chapter 5 investigates the potential and limitations of a traditional method (i.e. regional envelope curve) and a new method (i.e. physical model cascade) for flood mapping in a large river: the Blue Nile in Africa. This chapter is an application of flood mapping in ungauged basin using SRTM topography and other techniques.

Chapter 6 summarize the potential provided by low-cost, space-borne data in inundation modelling and demonstrate the limitations of integrating those data into inundation modelling. Further research needed on this topic is also discussed.

# CHAPTER 2

# INUNDATION MODELLING OF A MEDIUM RIVER: SRTM

# TOPOGRAPHY AND ERS-2 FLOOD EXTENT

This chapter focuses on assessing the usefulness of SRTM topography and ERS-2
SAR imagery in supporting 2D hydraulic modelling of floods. In particular, flood
propagation and inundation modelling of a 10-km reach of the River Dee (UK) is
performed by using LISFLOOD-FP to simulate the December 2006 flood event.
Flood extent maps from satellite imagery (ERS-2 SAR) and hydrometric information
(downstream water levels) are used as evaluation data. Uncertainty analysis is carried
out within a Monte Carlo framework using the roughness coefficients and downstream
water surface slope as free parameters. The results of this study show: (1) the
potential and limitations of SRTM topographic data in flood inundation modelling; (2)
the value of downstream water levels in constraining uncertainty in hydraulic model
of floods; (3) the impact of setting a water surface slope as downstream boundary on
the results of the hydraulic model (e.g. predictions of water levels and flood extent).

## 2.1  INTRODUCTION

Recent advances in airborne and satellite remote sensing allow the parameterization, calibration and validation of flood inundation models in a distributed manner (Bates 2004). Hydraulic models are usually tested on flood extent data (e.g. Matgen et al., 2007; Pappenberger et al., 2007; Neal et al., 2013) rather than water level or flow data at particular points as the models may not perform well at the locations away from the gauged points (Bates et al., 2004). As pointed out by Pappenberger et al., (2007), data used to constrain model parameter uncertainty should be consistent with the modelling purpose. For example, models are better to be conditioned on flood extent data if the goal is predicting flood prone areas, while high water marks are preferable if the purpose is estimating design flood profiles (Brandimarte and Di Baldassarre 2012). The reason is perhaps that a model can simulate flood extent well does not necessarily do well in water level in the same event, given the complexity of inundation modelling system.

Yet, the use of flood extent data can sometimes be difficult to distinguish between different model parameterization when the flood extent is not sensitive to changes in water level. In addition, flood extents from satellite flood images are usually difficult to obtain. As a matter of fact, the overpass frequency of the satellites which provides high resolution flood imagery is usually low (e.g. 35 days of repeat cycle for ERS2-SAR, Schumann et al., 2010) even though there are few products with low revisit time recently available (e.g. COSMO-SkyMed offers 12 h and 24 h revisit time, García-Pintado et al., 2013). This implies that finding a satellite image at the time of flooding may be difficult as flood duration time in small-medium catchment is usually shorter than the revisit time (Hunter et al., 2007; Schumann et al., 2010). Hydrometric data such as water levels are relative easier to find. They have a high temporal frequency, but are unavoidably sparse in space (Di Baldassarre et al., 2011).

Some scientists have explored the use of different types of datasets to constrain uncertainty in inundation models. For example, Horritt and Bates (2002) tested three hydraulic codes on a 60-km reach of the River Severn, UK using independent hydrometric and satellite data for model calibration. They found all models are capable of reproducing inundation extent and flood wave travel time to the similar level of accuracy at optimal simulation. However, the predictions of inundation extent are in some cases poor when hydrometric data are used for model calibration. Hunter et al., (2005) calibrated an inundation model against flood images, downstream stage and discharge hydrographs on a 35-km reach of the River Meuse, the Netherlands.

They found that the evaluation of internal predictions of stage also offer considerable potential for reducing uncertainty over effective parameter specification.

In hydraulic modelling, the normal depth (calculated from the water surface slope) is often used as downstream boundary condition. The water surface slope is normally unknown and is estimated as the average bed slope under the assumption of a Manning's type relationship between water level and discharge at the downstream end of the river reach. The results of flood inundation models (e.g. water levels, inundation extent) are affected by this assumption, especially when backwater effects are significant. Samuels (1989) proved the practical use of Equation (2.1) to calculate the backwater length, $L$, for engineering applications:

$$L = \frac{0.7D}{S_0} \qquad\qquad (2.1)$$

where $D$ is the bankfull depth of the channel and $S_0$ is the bed slope. Only a few studies (e.g. Wang et al., 2005; Schumann et al., 2008) have investigated the impact of assuming a certain water surface slope as downstream boundary conditions on the results of 2D hydraulic models, such as inundation extent and water level. Those impacts can be substantially reduced by extending the model domain and placing the downstream water level sufficiently far away from the points of interest. However, this is not always feasible.

As discussed above, the value of SRTM topography in supporting two-dimensional (2D) flood inundation modelling remains largely unexplored, particularly for medium-small sized (with width smaller than 100 m) rivers. In this context, the aim of this chapter is twofold: (i) explore the potential and limitations of SRTM data in supporting the 2D hydraulic modelling of floods, (ii) examine the sensitivity of 2D hydraulic models on the water surface slope used as downstream boundary as well as the associated value of downstream water levels in constraining uncertainty of flood extent prediction.

## 2.2   STUDY SITE AND DATA AVAILABILITY

The study is carried out on a river system including: (i) the 10-km reach of the River Dee, between Farndon and Iron Bridge, two gauging stations of the Environment Agency of England and Wales (hereafter called the EA); and (ii) the 8-km reach of the River Alyn, between the EA gauging station of Pont-y-Capel and the confluence to the River Dee (Figure 2.1). A high resolution (2 m) LiDAR DEM of this test site is derived by the EA. Surface artefacts such as vegetation and buildings are removed from the raw LiDAR data. The EA also conducts a channel bathymetry ground survey of 36 cross-sections which are incorporated with the LiDAR data on floodplain. Hereafter, this hybrid high resolution DEM is called LiDAR DEM.

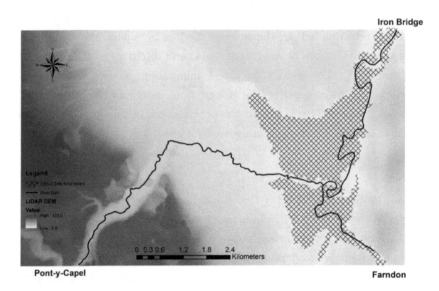

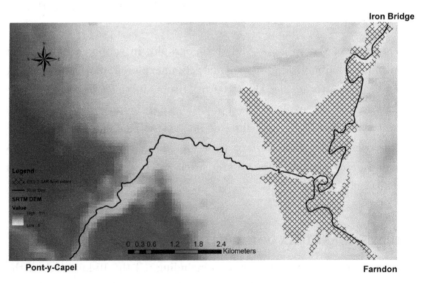

***Figure 2.1*** *River Dee between Farndon and Iron Bridge and River Alyn from Pont-y-Capel (black lines); Flood extent of 2006 event from ERS2-SAR flood image (crosshatch); LiDAR DEM (upper panel); SRTM DEM (lower panel)*

Another DEM of the study site is derived from the SRTM data post-processed by the Consortium for Spatial Information of the Consultative Group for International

Agricultural Research (CGIAR-CSI), e.g. fills in the no-data holes in the raw SRTM data (Jarvis et al., 2008). The SRTM DEM of the study area is reprojected into 75 m resolution with no speckles and surface artefacts removed. The two DEMs are strongly different, not only in terms of resolution (2 m versus 75 m), but also in terms of accuracy: the vertical accuracy of LiDAR data was of around 10 cm, while that of SRTM in Europe was found around 6 m (Rodríguez et al., 2006).

In December 2006, the River Dee experienced a low magnitude flood event (with the return period about 2 years). In this period, a high-resolution satellite image (ERS-2 SAR, see in Figure 2.1) was acquired. The ERS-2 SAR image is characterized by a pixel size of 12.5 m and a ground resolution of approximately 25 m. The satellite image was processed by using visual interpretation procedure to derive a flood extent map (Schumann et al., 2009; Di Baldassarre et al., 2010). We reproject this flood extent map into the 20 m and 75 m resolution for evaluating the LiDAR-based and SRTM-based models.

## 2.3  HYDRAULIC MODELLING

The LISFLOOD-FP (Bates et al., 2010) 2D hydraulic model is used to simulate the flood event in 2006. The main channel widths of River Dee and Alyn are on average 30 m and 12 m, respectively, which are much smaller than SRTM DEM cell resolution (i.e. 75 m). Therefore, the subgrid approach of LISFLOOD-FP (Neal et al., 2012), which can represent 1D channels with widths below the grid resolution, is applied for the SRTM-based model. However, computational cost can still be high for fine resolution (e.g. 1-10 m) grids. Therefore, the 2 m LiDAR DEM is aggregated into 20 m resolution to reduce the model computational time. The key topographic features such as embankments are manually identified in the aggregated DEM.

The channel bed elevation of SRTM topography is found to be overall overestimated in the study area. This is due to the fact that radar wave cannot penetrate water surface to detect the channel bed elevation and the channel is typically smaller than an SRTM pixel. The overestimated bed elevation would not only hamper the water level simulation in the main channel, it also affects flood extent prediction as the channel conveyance would most likely be decreased. Therefore, we improve the SRTM channel bed elevation by using the boat survey data. However, the combination of boat surveyed channel bed elevation and overestimated SRTM floodplain topography results in a deep channel depth (around 8 to 10 m). As one of the main purposes of inundation modelling is to predict the flood extent correctly, we use the surveyed channel depth (bank elevation subtract bed elevation) to replace the SRTM channel depth rather than directly replacing SRTM bed elevation by the surveyed bed elevation.

Two hydraulic models (LiDAR and SRTM-based) are built to simulate 2006 year flood event. The observed discharge hydrograph starting on 6 of December, 2006 at 11:00 h (around 144 h before the satellite overpass) is used as upstream boundary condition. A normal depth with the water surface slope (estimated as the average bed slope) is applied as the downstream boundary condition.

In this study, two types of observations are available for the model evaluation: (1) spatially distributed binary flood extent, (2) at-a-point time series of flood water levels (Iron Bridge). The simulated inundation areas are compared to the observed flood extent map (derived from the ERS2-SAR satellite imagery, Figure 2.1) using the performance measure, $F$ (Aronica et al., 2002; Horritt et al., 2007, Equation 1.5). As assumed in previous studies (e.g. Aronica et al., 2002; Pappenberger et al., 2007), only the cells with a simulated inundation depth greater than 20 cm are considered as flooded.

The evaluation of the simulated downstream water levels is conducted by using the observed time series of flood water levels at the downstream end of the river reach. The calibration focuses on the peak hours of the water level hydrograph, starting on 7 of December, 2006 at 4:00 am and ending 127 hours later, which is also the time of the satellite overpass. The root mean square error ($RMSE$) is used to evaluate model errors for both LiDAR and SRTM-based model.

## 2.4 THE EFFECT OF TOPOGRAPHY RESOLUTION

In order to better distinguish between the impact of the resolution and the accuracy of topographical input data, we first conduct a numerical experiment to isolate the resolution effect: the LiDAR DEM is aggregated into 80 m resolution, which is similar to the resolution of SRTM (i.e. 75 m). Then the subgrid LiDAR-based model (80 m of resolution) in which the channel has the same width, friction and bed elevation to the LiDAR-based model (20 m of resolution) is built. The SRTM-based model (75 m of resolution) is used for comparison. The other model parameters among three models are kept identical.

The flood extents simulated by three models are compared to the flood extent derived from the ERS2-SAR image. The value of performance measure, $F$, is shown in Table 2.1. The coarse resolution LiDAR-based model performs slightly worse than the high resolution one (with 0.010 difference in terms of $F$), while the performance is much higher than the SRTM-based model (with 0.271 difference in terms of $F$). This shows that coarse resolution LiDAR-based model can simulate the flood extent equally well as the high resolution LiDAR-based model. The coarse resolution does not degrade the model performance whereas the vertical accuracy of floodplain cells might play an important role. Thus, we focus on the effect of DEM vertical accuracy on flood extent and downstream water level predictions in the following experiments.

**Table 2.1** *The effect of spatial resolution: comparison of three floodplain models*

| Performance Measure | LiDAR-based model (80m) | LiDAR-based model (20m) | SRTM-based model (75m) |
|---|---|---|---|
| $F$ | 0.781 | 0.791 | 0.51 |

## 2.5 UNCERTAINTY ANALYSIS WITHIN A MONTE CARLO FRAMEWORK

In order to investigate the usefulness of SRTM data to support hydraulic modelling, the effects of topography uncertainty are evaluated within the Generalized Likelihood Uncertainty Estimation (GLUE, Beven and Binley 1992) framework. GLUE is a simple and pragmatic methodology, which uses Monte Carlo simulations to produce parameter distributions and uncertainty bounds conditioned on available data. GLUE has been widely used in environmental modelling (e.g. Aronica et al., 1998; Romanowicz and Beven 1998; Beven and Freer 2001). It is worth noting that a number of authors (e.g. Montanari 2005; Mantovan and Todini 2006; Stedinger et al., 2008) have shown that the GLUE methodology does not formally follow the Bayesian approach in estimating the posterior probabilities of parameters and the output distribution. Also, there are a number of subjective decisions to be made in GLUE; e.g. the priori distribution and feasible range of each parameter, (generalized) likelihood function for model evaluation, the threshold between behavioural and non-behavioural simulations. It is therefore necessary to clarify each decision to be transparent and unambiguous.

The assumed ranges of parameters do have an influence on resulting uncertainties (Aronica et al., 1998). Thus, large parameter ranges, which cover the extreme feasible values, are used to overcome the potential issue of subjective choice (e.g. Aronica et al., 1998). Therefore, we keep the roughness parameter range sufficiently large: both Manning`s channel and floodplain roughness coefficients are sampled randomly from the uniform distribution between 0.015 $m^{1/3}s^{-1}$ to 0.150 $m^{1/3}s^{-1}$ due to the lack of information regarding to the priori distribution as well as the feasible range of parameters. The similar parameter range was also used by Stephens et al., (2012) for the same study area. The range of average bed slopes is calculated from the two topographic data sets: the upper bound of the bed slope is calculated based on the reach from Fardon to Iron Bridge, while the lower bound is calculated based on the reach from the confluence to Iron Bridge for both LiDAR and SRTM-based model (Table 2.2).

*Table 2.2 Bed slopes calculated from LiDAR and SRTM topography*

| Models | Farndon to Iron Bridge | Confluence to Iron Bridge |
|---|---|---|
| LiDAR-based model | 0.0002 | 0.00002 |
| SRTM-based model | 0.0012 | 0.0002 |

The objective function (i.e. $F$ and $RMSE$) values for each parameter set (i.e. Manning's coefficients in channel ($n_{ch}$), Manning's coefficients on floodplain ($n_{fp}$) and water surface slope ($Sl$)) are calculated to derive the generalized likelihoods, which are positive values with the summation of 1 (Wagener et al., 2001). The likelihood measure can be used to weight each model realization. The behavioural models can be selected by rejecting the simulations that underperform a user-defined threshold or a percentage of simulations. In this study, the best 10% of the realizations are assumed as behavioural models and then used to produce probabilistic inundation maps. Given the simulation results for the $j$th computational cell of $w_{ij} = 1$ for wet and $w_{ij} = 0$ for dry, the probabilistic inundation map is produced using Equation 2.2:

$$M_j = \frac{\sum_i L_i w_{ij}}{\sum_i L_i}$$

$$(2.2)$$

Where $M_j$ indicates a weighted average flood state for the $j$th cell, $L_i$ is the likelihood weight assigned for each simulation $i$. The posterior parameter distributions (PPD) of both two models are plotted as well.

## 2.6  RESULTS AND DISCUSSION

We have conducted 1000 simulations for LiDAR and SRTM-based models within the GLUE framework (see above). Two performance measures are evaluated according to two types of observations. The dotty plots are generated to show the parameter uncertainty given alternative performance measures and data sets for both LiDAR and SRTM-based model.

Figure 2.2a and Figure 2.2b show that the performance measure of LiDAR-based model increases as the $n_{ch}$ and $n_{fp}$ increase when conditioned on flood extent data for the small $n_{ch}$ and $n_{fp}$ values (below 0.05). The performance measure begins to remain unchanged for the larger $n_{ch}$ and $n_{fp}$ values. It is found out that there is a tendency to generally underestimate the flood extent for smaller Manning's coefficients. The simulated inundation extent is, as expected, increasing when $n_{ch}$ and $n_{fp}$ are increasing.

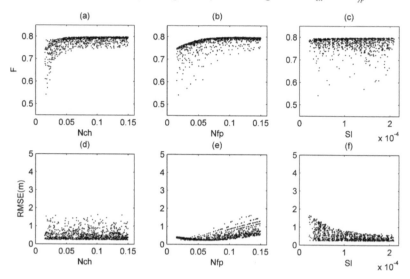

***Figure 2.2*** *(a), (b), (c): Performance measure F for LiDAR-based model conditioned on ERS2-SAR flood image; (d), (e), (f): Performance measure RMSE for LiDAR-based model conditioned on downstream water level.*

It is difficult to observe the optimal $Sl$ when conditioned on flood extent data as there are good simulations across the whole range of parameter values (Figure 2.2c). The flood extent is affected by the back water effect and therefore related to the downstream water slope $Sl$. However, the influence of backwater to flood extent in this case is limited as the floodplain acts as a "valley-filling" case, whereby, once the valley is filled by flood water, increases of water depth do not lead to significant differences in flood extent (Hunter et al., 2005).

The sensitivity to parameters $n_{fp}$ and $Sl$ is assessed also by conditioning the model on downstream water levels (Figure 2.2e, 2.2f). The $RMSE$ is increasing when $n_{fp}$ is increasing with the optimal value around 0.05. The $RMSE$ is decreasing when $Sl$ is increasing (Figure 2.2f). The effects of the two parameters are compensated with each other (e.g. when $Sl$ is increasing, one can keep almost the same water level but increase $n_{fp}$). The sensitivity of $Sl$ is clearly visible conditioned on water level information, as expected. The predicted downstream water levels are strongly affected by the assumed water surface slope.

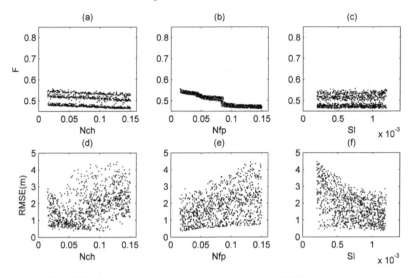

***Figure 2.3*** *(a), (b), (c): Performance measure F for SRTM-based model conditioned on ERS2-SAR flood image; (d), (e), (f): Performance measure RMSE for SRTM-based model conditioned on downstream water level.*

In Figure 2.3c, the average performance measure remains stationary with the change of $Sl$ when the SRTM-based model is conditioned on flood extent data. There is a clearly decreasing performance as $n_{fp}$ is increasing (Figure 2.3b). A similar trend can also be found for $n_{ch}$ (Figure 2.3a). The SRTM-based model essentially overestimate the flood extent (with a few under-prediction cells), even with the optimal parameter sets. This is shown by the fact that the performance measure ($F$) keeps dropping while $n_{ch}$ and $n_{fp}$ are increasing due to the fact that the simulated inundation extent keeps increasing, which result in more over-prediction.

The sensitivity of SRTM-based model conditioned on downstream water level (Figure 2.3d, 2.3e, 2.3f) is overall larger than LiDAR-based model. The realizations with high performance (low $RMSE$) are more concentrated in an area with the small values of $n_{ch}$ and $n_{fp}$ rather than the high ones. Similarly to the LiDAR-based model, the SRTM-based model performs better with larger $Sl$ values (Figure 2.2f, Figure 2.3f).

Figure 2.4b and 2.4e show that, when conditioned on downstream water levels, the best realizations are obtained with small Manning`s floodplain roughness values (around 0.05), whereas high performance realizations are found for higher Manning`s floodplain values, when the model is conditioned on flood extent. The posterior parameter distribution for water surface slope, when conditioned on flood extent, and

Manning`s channel coefficients, when conditioned on water level, are found nearly uniform distributed for the LiDAR-based model (Figure 2.4c, 2.4d). If all the realizations are taken as behavioural, we might conclude that the simulations conditioned on hydrometric data (i.e. water level time series) may not predict the flood extent properly for LiDAR-based model (Figure 2.4a to 2.4f). However, the performances are different after the rejection of non-behavioural simulations. Figure 2.5 (upper panel) shows the flood extent predicted by the best 10% simulations, conditioned on downstream water levels of which the average performance measure ($F$) is 0.771, given the best $F$ among all 1000 simulations is 0.799 (see Table 2.3).

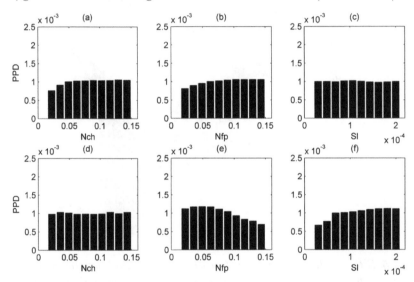

***Figure 2.4*** *(a), (b), (c): Posterior Parameter Distribution for LiDAR-based model conditioned on ERS2-SAR flood image; (d), (e), (f): Posterior Parameter Distribution for LiDAR-based model conditioned on downstream water level.*

In Figure 2.6, the posterior parameter distribution shows the performance of parameters for SRTM-based model is rather similar between the two performance measures (i.e. $F$ and $RMSE$). This indicates that it might get relatively satisfactory predictions of flood extent when the SRTM-based model is conditioned on water level data. It also shows SRTM-based model might be more flexible in conditioning on different data sets than LiDAR-based model. The model performances after the rejection of non-behavioural simulations are also shown in Figure 2.5 (lower panel). The probabilistic inundation map of the best 10% simulations conditioned on water level of which the average performance measure ($F$) is 0.524, compare with the best $F$ of 0.557 among all 1000 simulations (Table 2.3).

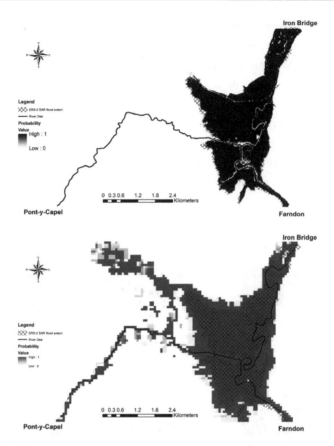

**Figure 2.5** *ERS2-SAR flood imagery (crosshatch) and Probabilistic inundation map of 2006 event (from black, 1, to white, 0) with behavioural simulations conditioned on downstream water level: LiDAR-based model (upper panel); SRTM-based model (lower panel).*

Figure 2.2 and Figure 2.3 show that the LiDAR-based model performs better than the SRTM-based model in predicting flood extent as well as the downstream water level. This shows how the performance of hydraulic models can be affected by topographic errors. The prediction of downstream water levels shows a mean *RMSE* of 1.853 m for SRTM-based model and 0.504 m for the LiDAR-based model. Considering the predicted water level is obviously affected by the channel bed elevation, the poor performance of the SRTM-based model is expected. On the other hand, the mean $F$ of the best 10% realizations conditioned on ERS2-SAR is 0.543 for SRTM-based model and 0.797 for LiDAR-based model (Table 2.3). Despite this large difference, getting a performance above 50% in simulating flood extent is a reasonably good result for using SRTM topography to support the hydraulic modelling of a small-medium sized river.

**Table 2.3** *Performance of behavioural LiDAR and SRTM-based models*

| Models | Average of best 10% simulations conditioned on water stage ($F$) | Average of best 10% simulations conditioned on ERS2-SAR ($F$) | Best of 1000 simulations ($F$) |
|---|---|---|---|
| LiDAR-based model | 0.771 | 0.797 | 0.799 |
| SRTM-based model | 0.524 | 0.543 | 0.557 |

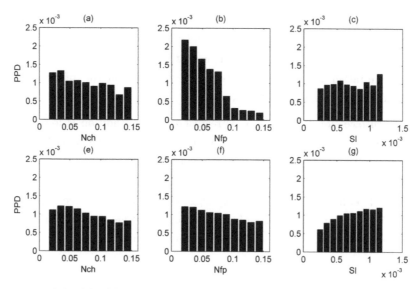

**Figure 2.6** *(a), (b), (c): Posterior Parameter Distribution for SRTM-based model conditioned on ERS2-SAR flood image; (d), (e), (f): Posterior Parameter Distribution for SRTM-based model conditioned on downstream water level*

## 2.7 CONCLUSIONS

This chapter presents an evaluation of the potential usefulness of SRTM topography in supporting models predicting flood extent as well as downstream water levels, by taking into account parameter uncertainty within a GLUE framework. The topographic uncertainty is estimated by comparing the SRTM-based model to a model based on high resolution topography (i.e. LiDAR plus channel survey). The ERS2-SAR flood imagery and downstream time series of water levels of the 2006 flood event are used to constrain model uncertainty. Roughness coefficients in channel and floodplain as well as the water surface slope are sampled uniformly within their parameter space. The effect of water surface slope in affecting flood extent and downstream water levels is quantified. The ability of a 2D flood inundation model conditioned on water level to simulate flood extent is also evaluated.

The SRTM-based model performs poorly for the downstream water level predictions, but it captures the majority of the inundation patterns. In addition, similar optimal parameters for the SRTM-based model conditioned on flood extent or water level are encouraging. However, to generalise these findings, SRTM data should be tested on more case studies.

It is also shown that the optimal parameters are rather different when the LiDAR-based model is conditioned on either the flood extent or water levels. However, when behavioural simulations are conditioned on water level, predictions of flood extent prediction are rather good. This is likely due to the fact that the differences in water levels do not imply changes in flood extent.

The water surface slope used as downstream boundary condition is found to have a negligible impact on flood extent predictions with the LiDAR-based model and a limited impact on flood extent predictions with the SRTM-based model. In contrast, the downstream water surface slope is found to significantly affect water level predictions of both models. This finding suggests that water surface slope has to be selected with caution when one of the purposes of the hydraulic model is the prediction of downstream water levels and design flood profiles.

# CHAPTER 3

## INUNDATION MODELLING OF A MEDIUM-TO-LARGE RIVER:

## SRTM TOPOGRAPHY AND ENVISAT FLOOD EXTENT

This chapter explore the value of SRTM topography and ENVISAT SAR image to support 1D flood inundation modelling under parameter and inflow uncertainty. The study is performed on a 98 km reach of the River Po in northern Italy. The comparison between a hydraulic model based on high-quality topography (i.e. LiDAR) and one based on SRTM topography was carried out by explicitly considering other sources of uncertainty (besides topography inaccuracy) that unavoidably affect hydraulic modelling, such as parameter and inflow uncertainties. In addition, the usefulness of SRTM topography in 1D hydraulic modelling is also explored in a model calibration-evaluation approach using two independent flood events. The results of this study show that the differences between the high-resolution topography-based model and the SRTM-based model are significant, but within the accuracy that is typically associated with large-scale flood studies.

## 3.1  INTRODUCTION

Results of hydraulic modelling are affected by many sources of uncertainty: model structure, model parameters, topography data and boundary conditions (e.g. Aronica et al., 1998; Romanowicz and Beven 1998, 2003; Bates et al., 2004; Pappenberger et al., 2005; Solomatine and Shrestha 2009). To better estimate the overall uncertainty, the uncertainty assessment exercises should be carried out throughout the modelling process, starting from the very beginning, rather than be added after the completion of the modelling work (e.g. Refsgaard et al., 2007; Merwade 2008; Di Baldassarre et al., 2010). In particular, the inflow (i.e. design flood) and the model parameter uncertainties were found to often be the most significant sources of uncertainty in one-dimensional (1D) hydraulic modelling for estimating design flood profiles (flood elevations at cross sections computed from design flood) (e.g. Pappenberger et al., 2008). Specifically, to construct inflow data for hydraulic models, the design flood estimation via flood frequency analysis is in turn affected by other sources of uncertainty, such as observation errors (e.g. rating curves), limited sample size and selection of the distribution model (Di Baldassarre et al., 2009a). In addition, more recently, Moya Quiroga et al., (2012) showed that topographic uncertainty has a considerable influence on flood extent by using a simplified sampling technique of changing the cell elevations independently.

Given that exploring low-cost space-borne data to support hydraulic modelling is the main objective of this thesis, in this chapter, the impact of using SRTM as the topographical input on the performance of flood inundation models is analysed by explicitly considering the other sources of uncertainty (such as parameters and inflow) that would unavoidably affect any hydraulic modelling exercise. The uncertainty introduced by topographic inaccuracy is therefore analysed in the following perspective: to what extent do the topographic errors affect the floodplain model results? And is this significant in view of the other sources of uncertainty (inflow, parameters) that are intrinsic to any modelling exercise?

Given its short revisit time and low cost, ENVISAT ASAR coarse resolution satellite imagery was used in many hydraulic modelling studies (e.g. Prestininzi et al., 2010; Pappenberger et al., 2007; Tarpanelli et al., 2013). However, ENVISAT imagery has rarely been used as calibration or evaluation data for hydraulic models built based on global DEM (such as SRTM topography). Therefore, the value of integration of coarse resolution imagery and global topography remains largely unexplored.

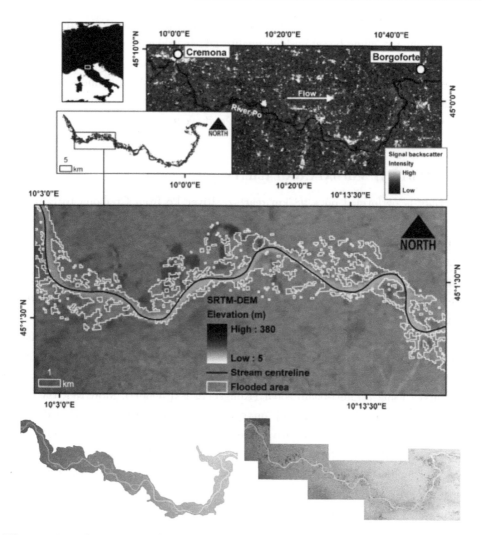

***Figure 3.1*** *Study site: ASAR image of the River Po between Cremona and Borgoforte (Italy) (upper panel, extracted from Di Baldassarre et al., 2009); SRTM-DEM, stream centreline and derived flooded area (middle panel, extracted from Di Baldassarre et al., 2009); LiDAR DEM (lower-left panel), SRTM DEM (lower-right panel)*

To this end, design flood profiles are estimated by using hydraulic models built based on SRTM topography. These design flood profiles are then compared to the ones provided by a hydraulic model based on accurate and precise topography (i.e. combination of LiDAR survey, multi-beam bathymetry and cross sections). This comparison is made in view of the associated uncertainty, which is estimated by following a simple and pragmatic approach proposed by Brandimarte and Di

Baldassarre (2012). They estimated the uncertainty in predicting design flood profiles via hydraulic modelling based on high-quality DEMs, while we develop a SRTM-based model and compare it with the LiDAR-based model and use a similar Monte Carlo-based approach to estimate uncertainty. In addition, the usefulness of SRTM topography in 1D hydraulic modelling is also explored in a model calibration-evaluation approach using two independent flood events.

## 3.2    STUDY SITE AND DATA AVAILABILITY

The study is performed on a 98 km reach of River Po from Cremona to Borgoforte (Figure 3.1). The Po River is the largest and longest river in Italy, when considering the drainage area of about 71,000 km$^2$ and its 650 km reach length. For the river reach studied, the main channel width varies from ~200 to ~300 m, while the river banks are confined by two lateral artificial levees with a width from 400 m to 4 km. A high-resolution 2 m DEM of this portion of River Po is constructed by the River Po Basin Authority. The DEM is constructed on the basis of data collected in 2005 during numerous flights from altitudes of approximately 1500 m, using two different laser scanners (3033 Optech ALTM and Toposys Falcon II). Below the water surface, channel bathymetry of the navigable portion was acquired during the same year by a boat survey using a multi-beam sonar (Kongsberg EM 3000D). Moreover, these data were complemented by the ground survey for 200 cross sections developed by the Interregional Authority of the River Po in 2005 (Castellarin et al., 2011). For convenience, and also because the major part of the DEM is constructed using the LiDAR technique, this DEM is called the LiDAR DEM in this study. A relatively coarse-resolution DEM of this portion of River Po based on the SRTM is also used for this study (Figure 3.1).

## 3.3    HYDRAULIC MODELLING

This study uses HEC-RAS (US Army Corps of Engineers 2001), a 1D hydraulic model that solves the De Saint-Venant equations with an algorithm based on the Preissmann implicit four-point finite scheme (Preissmann 1961). The model is used to simulate the flood profiles for the 98 km reach between Cremona and Borgoforte. In particular, two hydraulic models are built based on 88 cross sections including both the floodplain and the main channel. It is worth mentioning that the in-channel bathymetries of the LiDAR-based model come from the boat survey, while the in-channel bathymetries of the SRTM-based model are extracted from the raw SRTM

DEM without artificial manipulation. The locations of the cross sections for both LiDAR- and SRTM-based models are identical, while the cross-section bathymetries are significantly different, particularly in the main channel (Figure 3.2). Indeed, the use of SRTM topography for inundation modelling should be carefully justified due to the fact that radar waves cannot penetrate the water surface. Therefore, the in-channel bathymetries are poorly represented by SRTM topography in general. Conveniently, the SRTM topography data were obtained during an 11 day mission in February, which is winter time in River Po basin, meaning that the river water levels were relatively low and the floodplain was not inundated. The vegetation height and density during this winter time are also lower than in the summer. These give the opportunity to capture most of the in-channel bathymetries and provide relatively accurate floodplain topography.

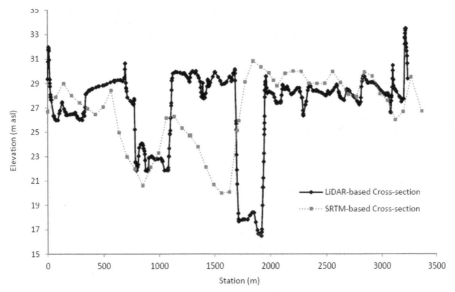

**Figure 3.2** *Cross-sectional comparisons for LiDAR-based model (solid line) and SRTM-based model (dashed line).*

Choosing HEC-RAS as the modelling tool are justified. Firstly, previous studies have shown that the unprotected floodplain, together with the main channel of this portion of the River Po between the artificial levees, can be treated as a compound 1D channel for a high-magnitude flood event (e.g. Castellarin et al., 2009; Di Baldassarre et al., 2009b; Brandimarte and Di Baldassarre 2012). Moreover, a 1D model is more efficient than a 2D model in terms of computational time, especially for large areas. It has been shown (Brandimarte and Di Baldassarre 2012) that for this test site, the computation of flood profiles via steady-state simulations provide results as accurate as the ones obtained via unsteady simulation because of the broad and flat hydrographs of this alluvial river, which leads to relatively little transient behaviour. Thus, a dynamic model is not needed here as the modelling purpose is to derive the maximum water stage levels, i.e. flood profiles. Moreover, it should be noted that Bates and De Roo (2000) demonstrated that, for example, for a particular 1-in-63 year flood event in River Meuse, the dynamic simulations are marginally less good predictors of inundation extent than the steady-state models. In addition, steady-flow routines are commonly adopted by regulatory agencies for floodplain mapping studies (Di Baldassarre et al., 2010).

**Table 3.1** *Mean absolute error for the optimal simulations of LiDAR-based model and SRTM-based model*

| Models | $N_{ch}$ | $N_{fp}$ | Mean Absolute Error (m) |
|---|---|---|---|
| LiDAR-based model | 0.04 | 0.06 | 0.27 |
| SRTM-based model | 0.03 | 0.05 | 0.56 |

## 3.4   MODEL CALIBRATION

In October 2000, the River Po experienced a significant flood event, with a peak discharge of about 11,850 m$^3$ s$^{-1}$. The return period was estimated at ~60 years (Maione et al., 2003). The two hydraulic models are calibrated by varying the Manning coefficients and comparing the simulated flood profiles to the high-water marks recorded after the October 2000 flood event (Coratza 2005), which is evaluated by using the mean absolute error, $\varepsilon$. Given the homogeneous characteristics of the river reach, the potentially distributed Manning`s n value are limited to one value for the channel and one for the floodplain (Di Baldassarre et al., 2009b). The high-water marks are appropriate to calibrate the model for the purpose of reproducing the flood profile. The model calibration is carried out by varying the Manning channel coefficient from 0.01 m$^{1/3}$ s to 0.06 m$^{1/3}$ s and the Manning floodplain coefficient from 0.03 m$^{1/3}$ s to 0.13 m$^{1/3}$ s, for both the LiDAR-based model and the SRTM-based model.

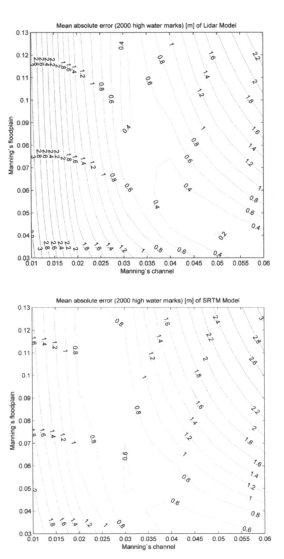

**Figure 3.3** *Model responses to changes in Manning coefficients: LiDAR-based model (upper panel) and SRTM-based model (lower panel)*

There is a clear tendency that the Manning channel coefficients are increasing, while the Manning floodplain coefficients are decreasing within areas with similar mean absolute errors for both LiDAR- and SRTM-based models (Figure 3.3). This is due to the fact that the increasing roughness on the channel is compensated for by the decreasing roughness on the floodplain. The shapes of the hyperbolic that contain the optimal values are similar between the two models. In addition, both models clearly show more sensitivity on Manning channel coefficients than Manning floodplain coefficients. The model sensitivity of the LiDAR-based model is higher than the SRTM-based model for small Manning channel coefficients. Figure 3.4 represents flood profiles of the best-fit model for both the LiDAR-based model and the SRTM-based model, which are selected by minimizing the mean absolute error and rejecting the simulations where the Manning floodplain coefficients are smaller than the Manning channel coefficients. The rejection criterion is based on the fact that there is a consensus that the channel roughness is typically smaller than that of the floodplain since water flows in the riverbed almost all the time and makes it smoother (Chow 1959). The mean absolute errors of the optimum simulation seem to indicate that both LiDAR and SRTM-based models can simulate the October 2000 flood profile reasonably well (Table 3.1), though the optimal Manning coefficients lie in different locations of the parameter space (Figure 3.3).

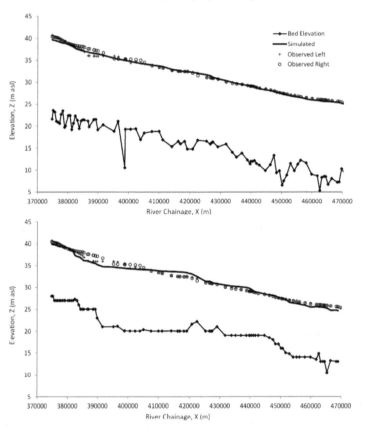

***Figure 3.4*** *Model calibration: observed left and right bank high-water marks and results of the best-fit model: LiDAR-based model (upper panel) and SRTM-based model (lower panel).*

## 3.5   MODEL EVALUATION

The River Po experienced a low-magnitude flood event (return period around 3 years) in June 2008. For this event, a coarse-resolution (~100 m) ENVISAT ASAR flood image was acquired and processed by Schumann et al., (2010). This ENVISAT ASAR flood image is used in this study to evaluate the two models. In particular, the inundation widths are derived to validate the best-fit LiDAR- and SRTM-based models. The two models with optimal Manning's coefficients in calibration are used to simulate the June 2008 flood, and the simulated inundation widths are compared to the ones derived by the SAR images. The validation of the two models (Figure 3.5) shows that they provide similar results. In particular, the mean absolute errors of the two models in reproducing the SAR inundations are found equal to 945 m for the LiDAR-based model and 862 m for the SRTM-based model. The difference between the two models is within the accuracy of the inundation width derived from the SAR image (around 150 m; e.g. Prestininzi et al., 2011). However, the mean absolute errors also indicate that both models cannot properly reproduce the inundation width of 2008 flood event.

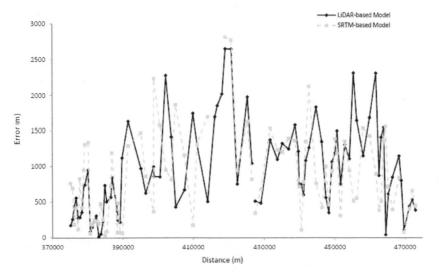

***Figure 3.5*** *Error of simulated and observed inundation widths of 2008 flood, LiDAR-based model (solide line) and SRTM-based model (dashed line).*

## 3.6   UNCERTAINTY ANALYSIS WITHIN A MONTE CARLO FRAMEWORK

The traditional approach to estimate design flood profiles is to feed the design flood into a calibrated model. In this case, the 1-in-200 year design flood (Q200 ~ 13,700 m$^3$ s$^{-1}$, estimated using Gumbel distribution) is simulated using the model calibrated against the October 2000 flood event (11,850 m$^3$ s$^{-1}$). The similar flood peaks magnitudes of the 1-in-200 year flood and the October 2000 flood indicate the profile predictions might be reliable, given that the high-magnitude floods are restricted by the two lateral banks of River Po, which means the flood propagation processes remain mainly 1D for this test site (e.g. Castellarin et al., 2009; Di Baldassarre et al., 2009b).

Recent literature has pointed out that single (deterministic) predictions of flood profiles, which use the 'best-fit' model and a single design flood estimation, might misrepresent the existence of uncertainties intrinsic to any hydraulic modelling exercise (Beven & Freer 2001; Bates et al., 2004; Beven 2006; Di Baldassarre et al., 2010). Therefore, the use of the probabilistic approach is increasingly recommended.

A rigorous approach for the estimation of all sources of uncertainty might be computationally heavy and requires strong assumptions of the nature of the errors. Hence, a simple pragmatic approach, based on GLUE (Beven and Binley 1992), is used here to estimate the design flood profile uncertainty. It is worth mentioning that GLUE is criticized by some researchers as it requires a number of subjective decisions and uses a non-standard notion of likelihood (Hunter et al., 2005; Mantovan and Todini 2006; Montanari 2007; Stedinger et al., 2008). At the same time, it is still a widely used approach to estimate uncertainty in hydrological modelling (e.g. Montanari 2005; Winsemius et al., 2009; Krueger et al., 2010). One aspect that, as we discovered, has an influence on many Monte Carlo related methods (such as GLUE) is the sampling strategy used (Kayastha et al., 2010), but such an analysis is out of the scope of this thesis, therefore it is left for future studies.

The sources of uncertainties estimated here are the inflow (i.e. design flood) and the model parameters. Topography uncertainty is also taken into account by simulating flood profiles using two models based on two different topographies. As mentioned above, the quality of the LiDAR DEM is high in terms of both resolution and vertical accuracy. The usefulness of SRTM topography is estimated by comparing the design flood profiles predicted by the SRTM-based model with the design flood profiles predicted by the LiDAR-based model, which is considered as a reference model. These profiles are compared by including an explicit representation of inflow uncertainty and model parameter uncertainty. It is assumed here that these sources of uncertainty prevail over the uncertainty induced by imprecise model structure, and this can be justified by the (aforementioned) satisfactory results obtained by previous studies in simulating high-magnitude events in this test site via HEC-RAS modelling (e.g. Castellarin et al., 2009; Di Baldassarre et al., 2009b).

### 3.6.1    Parameter uncertainty

The first modelling exercise is carried out by considering parameter uncertainty only, varying the Manning channel coefficient in the range 0.01–0.06 m$^{1/3}$ s$^{-1}$ (with an interval of 0.01) and the Manning floodplain coefficient in the range 0.03–0.13 m$^{1/3}$ s$^{-1}$ (with an interval of 0.01). A set of simulations satisfying a threshold criteria are selected from the sensitivity analysis (the same as the model calibration). In this case, all the simulations (66 in total) associated with a mean absolute error that is larger than 1 m are rejected. This (subjective) assumption is by considering the fact that the current policy in the Po River basin requires that levees should have at least 1 m of freeboard above the 200 year flood profile elevations (see Brandimarte and Di Baldassarre, 2012). In addition, the simulations when the Manning coefficients on the floodplain are smaller than those on the channel are rejected (see above). The models passing these filters are termed (following the GLUE method) 'behavioural', and are used to simulate the 1-in-200 flood event.

In GLUE, each behavioural simulation is associated with a rescaled likelihood weight, $W_i$, ranging from 0 to 1 within the framework of GLUE (Beven and Binley 1992). Many studies (e.g. Mantovan and Todini 2006; Beven et al., 2007, 2008; Stedinger et al., 2008) have pointed out that the likelihood function used in GLUE should be able to correctly represent the statistical sampling distribution of the data to make the prediction coherent and consistent. However, it should be noted that the choice of likelihood measure might greatly influence the resulting uncertainty intervals, and this choice must be made explicit so they can be the 'subject of discussion and justification' (Beven and Freer 2001). Here, we follow the previous studies in hydraulic modelling of floods (Bates et al., 2005; Hunter et al., 2005; Pappenberger et al., 2006). The likelihood weight, $L_i$, is expressed as a function of the measure of fit, $\varepsilon_i$, of the behavioural models:

$$L_i = \frac{\varepsilon_{max} - \varepsilon_i}{\varepsilon_{max} - \varepsilon_{min}} \qquad (3.1)$$

where $\varepsilon_{max}$ and $\varepsilon_{min}$ are the maximum and minimum values of the mean absolute error of the behavioural models. Then, the likelihood weights are rescaled to a cumulative sum of 1. Rescaled likelihood weights are calculated using:

$$W_i = \frac{L_i}{\sum_{i=1}^{n} L_i} \qquad (3.2)$$

This uncertainty analysis is implemented for both the LiDAR-based model and the SRTM-based model. Thus, Figure 3.6(upper panel) shows the 5th, 50th and 95th weighted percentiles, which represent the likelihood weighted uncertainty bounds, for both the LiDAR-based model and the SRTM-based model.

### 3.6.2    Inflow uncertainty

The second modelling exercise also considers the important source of uncertainty in hydraulic modelling, i.e. the estimation of the design flood. To this end, the 1-in-200 year flood is estimated by statistically inferring the time series of 42 annual maximum flows, recorded at the Cremona gauge station, which is the upstream end of the river reach under study.

In particular, five distribution functions commonly used in extreme value analysis in hydrology (i.e. lognormal (LN), three-parameter lognormal (LN3), exponential (E), Gumbel (EV1) and generalized Pareto (GP)) are fitted to the annual maximum flows using the method of moments. Five 1-in-200 year design floods are then obtained from these fitted distributions (Table 3.2). The shapes of the design hydrographs are not estimated because of the steady-state assumption already discussed. Thus, the additional uncertainties that might have been caused by the estimation of hydrograph shapes are avoided here. Hence, five simulations of the best-fit models are run for both the LiDAR- and the SRTM-based models. The 5th, 50th and 95th percentiles were computed. Comparisons are made between profiles for each percentile of the LiDAR-based model and the SRTM-based model (Figure 3.6 (middle panel)).

*Table 3.2* *1-in-200 year design flood estimation using five probabilistic models*

| Distribution | LN | LN3 | E | EV1 | GP |
|---|---|---|---|---|---|
| Discharge (m$^3$/s) | 13743 | 13254 | 15037 | 13695 | 12449 |

### 3.6.3    Combined uncertainty

The third experiment is carried out by combining model parameter and inflow uncertainties. Therefore, we run a total of 115 simulations by feeding 23 behavioural models using 5 design flood values generated from 5 distribution functions for the LiDAR-based model. For the SRTM-based model, we run a total of 120 simulations by feeding 24 behavioural models using the same 5 design flood values. The 5th, 50th and 95th weighted percentiles of the 1-in-200 year flood, which take into account the parameter and inflow uncertainties, were calculated for both the LiDAR-based model and the SRTM-based model. The comparisons of profiles from two models (LiDAR and SRTM) of the same percentile are also made (Figure 3.6 (lower panel)).

For the three experiments mentioned above, the chosen Manning coefficient parameter space is sufficiently large to cover the possible coefficient combinations for hydraulic modelling (Chow 1959). The sampling interval of 0.01 is reasonably small given by a sensitivity analysis carried out, which is to increase and decrease the sampling interval of floodplain Manning coefficients to 0.02 and 0.005, respectively. The results show that the uncertainty bounds for all three experiments have a limited variation. This means the uncertain model performance will not be affected much if the number of simulations is further increased.

## 3.7 RESULTS AND DISCUSSION

The bed elevations in the SRTM and LiDAR models are significantly different from each other (see in Figure 3.4). We also conduct a two-sample Kolmogorov–Smirnov test to quantitatively evaluate these differences. We test the hypothesis $H0: P1=P2$ that two samples come from the same distribution. The results show that it rejects the null hypothesis as the $p$-value $2.78 \times 10^{-6}$ is much smaller than 0.05, which is the default value of the level of significance. Therefore, the difference between two samples is significant enough, as they have different distributions. Figure 3.2 shows that the bed elevations differ by around 4 m, which definitely affects the main channel conveyance. Given that the cross-sectional profile of the main channel is significantly important for 1D hydraulic modelling, the following question may arise: why the modelling results are relatively close in terms of simulated water levels? In fact, the topography of the floodplain areas (see in Figure 3.2) of the two DEMs is not as different as that of the main channel (This complies with the aforementioned SRTM characteristics). Therefore, as the main channel and floodplain within the lateral embankments are treated as a whole cross section in this 1D HEC-RAS model, the total conveyance in flood conditions is not significantly different. In addition, the topographic uncertainties in the SRTM geometry are partly compensated for by the Manning coefficient during model calibration. All these factors result in similar simulated flood profiles for the LiDAR- and SRTM-based models, despite the fact that the 'conveyances' in the main channel are different.

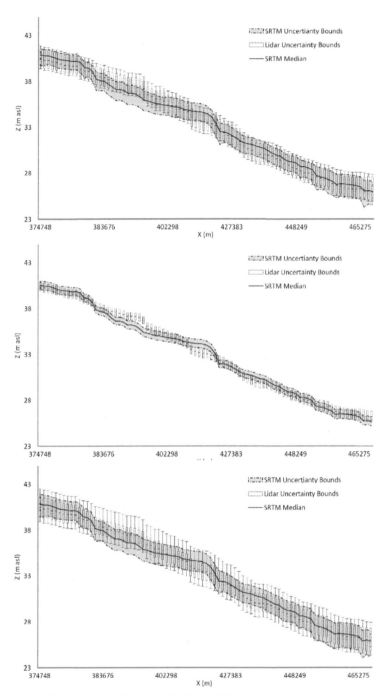

**Figure 3.6** *Uncertainty design flood profiles by considering: model parameter uncertainty only (upper panel); design flood uncertainty only (middle panel); and model parameter and design flood uncertainty (lower panel). LiDAR-based model*

*uncertainty bounds are shown in the boxplots (lower and upper quantiles are 5th and 95th percentiles). SRTM-based model uncertainty bounds are shown in the grey areas (dashed lines show 5th and 95th percentiles).*

Figure 3.6 shows that the uncertainty bounds vary according to the source of uncertainty under consideration. For the SRTM-based model, the difference between the 95% percentile and the median is 0.4–0.5 m if only inflow uncertainty is considered. This value increases to 0.6–1.0 m if only parameter uncertainty is considered. It continues to rise to 0.9–1.2 m if both parameter and inflow uncertainty are taken into account (Figure 3.6). Therefore, both sources of uncertainty contribute to the overall uncertainty. Moreover, the bounds of inflow uncertainties are relatively small. This is due to the fact that the differences of design flood estimations generated from 5 distributions are small (within ±10% of the 1-in-200 year flood estimated by the Gumbel distribution (EV1)).

**Table 3.3** *Difference of flood profiles (in terms of MAE) obtained from LiDAR-based model and SRTM-based model by considering uncertainties*

| Source of uncertainty | Percentile | *MAE* (m) |
|---|---|---|
| Parameter uncertainty | 5th percentile | 0.46 |
|  | 50th percentile | 0.47 |
|  | 95th percentile | 0.32 |
| Inflow uncertainty | 5th percentile | 0.41 |
|  | 50th percentile | 0.40 |
|  | 95th percentile | 0.39 |
| Combined uncertainty | 5th percentile | 0.47 |
|  | 50th percentile | 0.43 |
|  | 95th percentile | 0.36 |
| Mean |  | 0.41 |

The hydraulic model performance based on SRTM topography are evaluated when the LiDAR-based model results are taken as the 'truth', by comparing the flood profiles obtained from the LiDAR- and SRTM-based models for the three experiments (Table 3.3). The averaged mean absolute error of LiDAR- and SRTM-based model profiles in terms of 5th, 50th and 95th percentile, taking the major sources of uncertainties into account, is around 0.41 m. It is the same order of magnitude as the accuracy of the high-water marks (around 0.5 m; e.g. Neal et al.,

2009; Horritt et al., 2010). In addition, the current policy in the Po River basin requires that levees should have at least 1 m of freeboard above the 200 year flood profile elevations.

In this study, model performance evaluation is always based on water levels, as the calibration data of the October 2000 flood event are high-water marks. In addition, the estimation of the 1-in-200 flood profile is usually performed for dyke or levee design, which requires estimation of the flood water levels. In fact, the water depth differences for the LiDAR-based model and the SRTM-based model have up to 30% relative error for all the modelling practices because the SRTM DEM overestimates the topography elevation, particularly in parts of the main channel.

## 3.8  Conclusions

This chapter describes a study to quantitatively evaluate the value of the SRTM topography to support hydraulic modelling with the purpose of predicting design flood profiles when taking into account the main sources of uncertainty unavoidably associated with any modelling exercise. The outcomes of this study show that the prediction of 1-in-200 year flood profiles in the study area of River Po, using the HEC-RAS model based on SRTM topography data, are within the accuracy that is typically associated with large-scale flood studies, as long as the other sources of uncertainty are explicitly considered, despite the fact that some differences in water-level predictions (compared to the LiDAR-based model) are not negligible. Thus, the results indicate the added value of SRTM topography in supporting the prediction of design flood profiles in medium- to large-scale rivers, even though detailed studies for the ultimate design of hydraulic structures (e.g. bridges) do require more accurate modelling based on the best available topography. This study also investigates the uncertainty in hydraulic modelling due to uncertainty in the model parameters and design flood estimation. It is shown that both sources of uncertainty have a strong influence on the uncertainty of the model output.

The outcomes of this study are unavoidably associated with the particular nature of the case study. 1D modelling can, for example, be inappropriate in other cases, especially when the floodplain is not confined by two lateral artificial levees. Thus, future research will focus on the potential value of the SRTM-based hydraulic model tested on case studies with significant 2D inundation patterns. This study is also limited by the particular scale of the river. In smaller rivers, SRTM-based models are not expected to have the reasonable performance compared to LiDAR-based models presented in this study. Hence, the usefulness of the globally and freely available

SRTM topography data to support hydraulic modelling will be further tested by considering different scales and flood scenarios. In addition, in order to have a comprehensive evaluation of hydraulic models based on SRTM topographic data, different model calibration and evaluation approaches should be implemented according to the hydraulic model used, the available data and the source of uncertainty considered. Lastly, it should be noted that some subjective assumptions are made in estimating uncertainty via the Monte Carlo framework, and there have been no rigour tests carried out about the statistical validity of the conclusions, so these conclusions should be seen as the qualitative ones.

# CHAPTER 4

# INUNDATION MODELLING OF A LARGE RIVER: SRTM

# TOPOGRAPHY AND ENVISAT ALTIMETRY

This chapter aims to assess the value of SRTM topography and radar altimetry in supporting flood level predictions in data-poor areas. To this end, we build a hydraulic model of a 150 km reach of the Danube River by using SRTM topography as input data, and radar altimetry of the 2006 flood event as calibration data. The model is then used to simulate the 2007 flood event and evaluated against water levels measured in four stream gauge stations. Model evaluation allows the investigation of the usefulness and limitations of SRTM topography and radar altimetry in supporting hydraulic modelling of floods.

## 4.1 INTRODUCTION

Given that many areas in the world are effectively ungauged and the existing monitoring networks are under decline (Stokstad 1999), hydrometric data for hydraulic model building and calibration are also often unavailable. Satellite radar altimetry, which is typically used for oceanographic applications, may offer a great opportunity for the selection and calibration of hydraulic models in data-poor areas. Radar altimetry is also able to provide water level measurements in the continental environment, such as inland seas, lakes and rivers as demonstrated by the numerous studies in literature (Koblinsky et al., 1993; Birkett 1998; Getirana et al., 2009; Birkinshaw et al., 2010; Santos da Silva et al., 2010). The accuracy of altimeter water level time series is variable and ranges from 70 cm of *RMSE* for altimetry on board of Geodetic Satellite (GeoSat) (Koblinsky et al., 1993), to 30 cm for altimetry on board of Environmental Satellite (ENVISAT) (Santos da Silva et al., 2010). However, in some cases the discrepancies between altimetry and in-situ level measurements might be very high and are of the order of 2 m. This error depends on the type of the sensor used and the distance between the sensor and virtual station (VS), where radar satellite tracks intersect with a river reach and the ground gauged station. The high accuracy of altimetry data provided by the latest spatial missions and the convincing results obtained in the previous applications suggest that these data may also be used in the calibration of hydrological or hydrodynamic models (Getirana et al., 2009; Sun et al., 2010; Milzow et al., 2011; Pereira-Cardenal et al., 2011). For example, Wilson et al., (2007) presented a first attempt to simulate flood inundation extents as well as water levels in a 260-km reach of the Amazon at a large spatial scale over 22-month period by using SRTM topography, Japan Earth Resources Satellite (JERS)-1 images, gauge observations and satellite altimetry data. It was found that the *RMSE* between simulated water levels and the ones derived from satellite altimetry data was around 2 m. This error was mainly attributed to the uncertainty of topographic and altimetry data. However, the potential of radar altimetry for the calibration of hydraulic models in data-poor areas is still poorly explored, particularly when SRTM is used as input topographic data.

To this end, this chapter aims to analyse the ability of a hydraulic model, built and calibrated with freely available remote sensing data (i.e. SRTM topography and radar altimetry), in simulating flood water levels. Specifically, a hydraulic model of a 150 km reach of the Danube River is built using SRTM topography and calibrated against water levels derived from radar altimetry data of the 2006 flood event. The optimal model is then used to predict water levels of the 2007 flood event and evaluated by using ground water levels measured in internal cross sections.

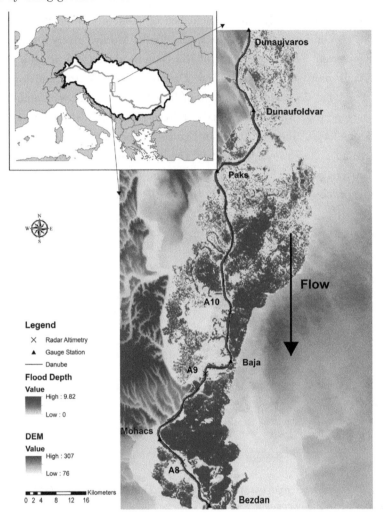

***Figure 4.1*** *Danube river between Dunaujvaros and Bezdan, A8, A9 and A10 are three radar altimetry virtual stations; simulated flood extent of 2007 event using optimal model in calibration (data from Jarvis et al., 2008, Sandvik 2008, European Environment Agency 2011)*

## 4.2   STUDY SITE AND DATA AVAILABILITY

The study is carried out on a 150-km reach of the Danube, between the stream gauging station of Dunaujvaros and the Global Runoff Data Centre (GRDC) station of Bezdan (Figure 4.1). The SRTM topography of the study area is projected into 80 m resolution with no speckles and surface artefacts removed.

In April 2006, an extremely rare coincidence of relatively large floods in the sub-basins of the Upper Danube, Tisza, Sava and Velika Morava rivers resulted in a 1-in-100 year flood event along this reach of the Danube river (with a peak discharge of 8400 m³/s at Dunaujvaros gauge station, Figure 2). The Central and Lower Danube were inundated by flood water due to melting snow and heavy rainfall (Wachter 2007). In this period, water levels derived from ENVISAT altimetry data at three virtual stations (A8, A9 and A10, see in Figure 4.1) are downloaded from the website of the River and Lake Project handled by the European Space Agency (ESA) and De Montfort University (http://earth.esa.int/riverandlake). At each VS three water level values are retrieved every 35 days for in total 9 observations (Figure 4.2).

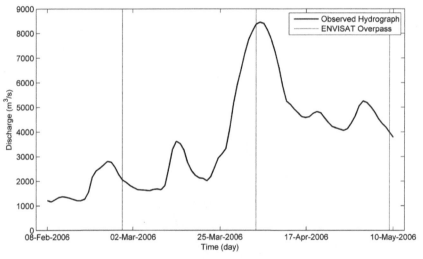

***Figure 4.2*** *Discharge hydrograph observed at Dunaujvaros of 2006 event and the timing of ENVISAT overpass*

In September 2007, Danube experienced another flood event with the peak discharge around 6400 m³/s at Dunaujvaros gauge station. River discharges and water levels of the 2006 and 2007 flood events were observed at this gauge station as well as at the other four gauge stations: Dunafoldvar, Paks, Baja and Mohacs (Figure 4.1, Table 4.1). In this study, a period of 91 days started from 8th of February to 10th of May is used for the 2006 event (Figure 4.2), while a period of 33 days started from 30th of August to 1st of October is used for 2007 event.

## 4.3  HYDRAULIC MODELLING

The LISFLOOD-FP (Bates et al., 2010) 2D hydraulic model is used to simulate the two flood events on Danube. We use an adaptive time step in LISFLOOD-FP. This implies that the time step is not defined by the modeller, but automatically computed by the model code to keep accuracy and numerical stability (Hunter et al., 2005). LISFLOOD-FP was largely used for flood inundation modelling and has provided positive results in numerous test sites (e.g. Neal et al., 2011, 2012a; Horritt and Bates, 2002).

*Table 4.1* Gauge Station characteristics in the study area

| Station Name | Distance to River Mouth (Black Sea, km) | Catchment Area (Upstream of Station, km$^2$) |
|---|---|---|
| Dunaujvaros | 1580.6 | 188273 |
| Dunafoldvar | 1560.6 | 188700 |
| Paks | 1531.3 | 189092 |
| Baja | 1478.7 | 208282 |
| Mohacs | 1446.9 | 209064 |

The river width in this reach of the Danube is around 500 m, which is much bigger than SRTM DEM cell resolution (i.e. 80 m). As a result, the channel width along the reach can be easily identified from the SRTM DEM. The computational time is found long (around 40 hours on a dual-core 2.30 GHz machine) for a single simulation of the 2006 flood event, which lasts for 91 days, using a grid size of 80 m. Additionally, it is found out that the random noise in SRTM data tends to be independent in each cell and can be reduced linearly in proportion to $1/n$ by aggregating cells, where $n$ is the number of cells being aggregated (Rodríguez et al., 2006). Thus, the 80 m SRTM DEM is aggregated to 160 m to both decrease the computational time and reduce the noise.

The SRTM's SAR based interferometer technology cannot obtain the geometry of the river bed below the water surface. Instead, the water elevation in the Danube River at the time of the mission is measured (February 2000; Farr et al., 2007). Thus, the channel bed elevation is overestimated in SRTM data. As mentioned in Chapter 1, the scientific community has proposed different approaches to deal with this issue (e.g. Neal et al., 2012a; Patro et al., 2009). Patro et al., (2009) reduced the SRTM DEM-derived cross-section elevation values by 2.3 m, which was the average difference between the elevations of SRTM DEM and topographic maps, to simulate the river flows with limited available data of Mahanadi River in India. Biancamaria et al., (2009) tested the uncertainty in river depth of hydraulic modelling by using three constant river depth values (5 m, 10 m and 15 m) in Ob River of Siberia. More recently, Neal et al., (2012a) used hydraulic geometry relationships to correct SRTM data and approximate the river bed elevation in an 800 km reach of the river Niger in Mali. In particular, they used a power law relationship between channel width and depth following the relationships of Leopold and Maddock (1953).

In this study, we follow a simple approach to correct the SRTM data, by lowering the bed elevation by $D$ meters, and introduce $D$ as a parameter of the model. We extract the river bed elevation (actually water surface elevation) within the bank from SRTM DEM, subtract those elevations by $D$ meters and 'burn' the new bed elevation into SRTM DEM. The 'new' SRTM DEMs (with various bed elevations depending on $D$ value) are used in hydraulic modelling. To avoid subjective assumptions and potential over-fitting, $D$ is set as a single, homogenous parameter along the entire river reach.

In hydraulic modelling, the calibration of hydraulic models is typically done by adjusting roughness coefficients in the channel and floodplain areas (e.g. Aronica et al., 2002; Pappenberger et al., 2005). In this study, given that the parameter $D$ has an impact on model results similar to channel roughness ($n_{ch}$). They both alter the channel conveyance. Thus, we follow a parsimonious approach and set the value of $D$ and the floodplain roughness coefficient ($n_{fp}$) as the only two parameters.

## 4.4  MODEL CALIBRATION

The 2006 event is the calibration event simulated by the LISFLOOD-FP model. The observed discharge hydrographs at Dunaujvaros gauge stations are used as upstream boundary conditions. The downstream boundary conditions are defined by the normal depth assumption: the water surface slope (if it is unknown) is estimated as the average bed slope under the assumption of a Manning's type relationship between water stage and discharge at the downstream end of the river reach. For both model calibration and evaluation, the water surface slope is approximated using the average slope derived from SRTM topographic data, which, in spite of the aforementioned inaccuracies, was demonstrated to provide a proper estimation of water surface slopes (e.g. Schumann et al., 2010; Alsdorf et al., 2007). Anyhow, the four stream gauging stations are far (e.g. Mohacs is 21 km away) from the downstream end so that the simulated water levels would not be affected by the imposed slope. No differences in simulated water levels in the four stream gauging stations have been found while employing various slope values in preliminary runs. For both model calibration and evaluation, the initial conditions are set as dry bed as no significant differences have been found if an arbitrary steady discharge is considered as initial condition.

As mentioned, the calibration of the LISFLOOD-FP model focuses on the identification of the two free parameters: the Manning's roughness coefficient of the floodplain, $n_{fp}$, and the SRTM adjusting factor for the river bed, $D$. The simulations are done by varying $D$ in the range between 0 and 7 m and $n_{fp}$ from 0.05 to 0.23 sm$^{-1/3}$. The main channel roughness coefficient, $n_{ch}$ is fixed to 0.04 sm$^{-1/3}$ for all the simulations. This value is set based on user experiences of previous studies in LISFLOOD-FP model calibration (e.g. Horritt et al., 2001; 2002) and the physical characteristic of this river each. The performance of the hydraulic model is estimated in terms of mean absolute error, $MAE$, calculated by comparing the water levels observed in three VSs with the ones simulated by the model. The minimization of $MAE$ allows estimating the optimum values of $n_{fp}$ and $D$.

Random search methods have been shown to have a potential to solve large-scale problems efficiently in a way that is not possible for deterministic algorithms. More advanced randomized search methods such as Genetic Algorithms (GA, Holland 1992, Goldberg 1989) and Adaptive cluster covering (ACCO, Solomatine 1995, 1999) can certainly improve the efficiency of the calibration. However, a simple algorithm such as (deterministic) grid search can already provide accurate results due to several reasons. First of all, the sampling dimensions are limited (i.e. two) and the initial searching range is relatively narrow in this case. Thus, the number of required samples grows exponentially in dimension is not an issue here. On the other hand, uncertainty induced by using sampling techniques in the calibration might not be comparable to that induced by other sources in flood modelling (e.g. observed data, model structure). Preliminary runs show that several simulations provide very similar performance near the optimal (less than 0.01 m difference in terms $MAE$, which is quite negligible for the flood issue in such a big river), therefore, the sampling interval used here is sufficiently small to identify the optimal. This allows us to find the optimal parameter set by using grid search method rather quickly. Given each simulation takes about 250 minutes on a duel core 2.3G Hz machine, an advanced random search methods with large number of runs would not be more efficient than the grid search method.

Equifinality is shown here in the calibration: the darkest irregular polygon in Figure 4.3 (indicates same $MAE$) covers $n_{fp}$ ranging from 0.14 sm$^{-1/3}$ to 0.22 sm$^{-1/3}$, while $D$ varies from 3 m to 5m. The issue of Euqifinality is due to the effect of parameter compensation. Here we choose the optimal parameter set based on its lowest $MAE$ as well as the reasonable physical meaning of parameters. For example, as the second largest river in Europe, the depth of Danube of this reach is normally deeper than 4 m (Andreadis et al., 2013).

## 4.5  MODEL EVALUATION

The calibrated model with optimum parameter sets is used to simulate the 2007 flood event. The observed discharge hydrographs in 2007 at Dunaujvaros gauge stations are used as upstream boundary conditions. The optimum model is evaluated using the flood water levels observed at the four stream gauging stations along the reach: Dunafoldvar, Paks, Baja and Mohacs. For each station, the $MAE$ between observed and simulated water level hydrographs is calculated.

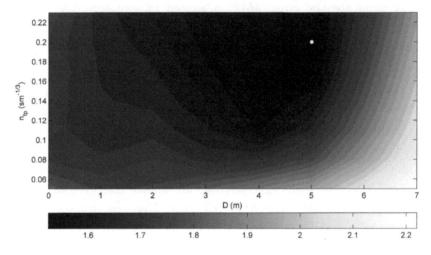

**Figure 4.3** Model Calibration: Contour map of MAE (m) conditioned on radar altimetry by varying D (m) and Manning`s floodplain coefficients ($n_{fp}$). The white dot indicates the 'best fit' model parameterization.

## 4.6  RESULTS AND DISCUSSION

Figure 4.3 shows the calibration contour map in terms of $MAE$, which is calculated based on nine radar altimetry water levels (3 days in 3 VSs). Model performance generally increases when $n_{fp}$ is increasing, whereas the performance first increases and then decreases when $D$ is increasing. Flood water levels tend to be overestimated with high bed elevations ($D$ from 0 to 3 m), and underestimated with lower bed elevations ($D$ from 6 to 7 m). The optimal model (~1.53 m $MAE$) corresponds to $D$ equal to 5 m and $n_{fp}$ equal to 0.20 sm$^{-1/3}$.

The evaluation of the model shows that the peak water levels are underestimated at Baja and Mohacs, while the tails of the hydrograph are generally overestimated for the four gauge stations (Figure 4.4). The average $MAE$ at the four gauge stations of model evaluation is 1.37 m (Table 4.2). It is interesting to note that this value is similar to the value obtained in model calibration versus radar altimetry (1.53 m). Considering that the model was built with SRTM topography and calibrated with altimetry data, these results are encouraging for large-scale flood modelling in data-poor areas.

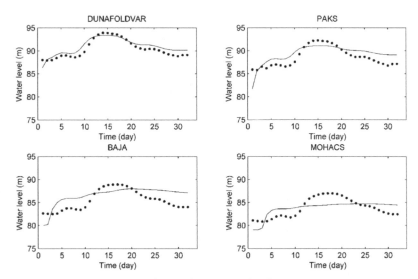

**Figure 4.4** *Model evaluation (2007 flood event): Simulated water levels hydrograph (line) and in situ water levels (dots) at the four gauging stations by using optimal model of the 2006 event*

These errors are likely due to the combination of: (1) inaccuracy of SRTM topography used to build the model; (2) water surface slope derived from altimetry used as downstream boundary condition in model calibration; (3) vertical errors in radar altimetry data used to calibrate the model; (4) other sources of uncertainty, e.g. model structure or inflow uncertainty (Di Baldassarre et al., 2012). For instance, considering the high magnitude of 2006 event (1-in-100 year), the model uncertainties which were compensated in model calibration by adjusting model parameters can impact model results when a relatively smaller flood event is simulated (see e.g. Di Baldassarre et al., 2011).

By analysing Figure 4.4, one can observe that the flood water levels in Dunaflodvar are generally well reproduced (with 0.75 m *MAE*, Table 4.2), although this station is close to the upstream boundary. Water levels in Parks and Baja are moderately well simulated with a limited underestimation of the flood peak, while water levels in Mohacs are poorly simulated. It is worth noting that in the model evaluation the simulated water levels tend to be less variable than the observed ones and the peaks are not modelled well (Figure 4.4). This effect is also exacerbated from upstream to downstream stations.

**Table 4.2** *Evaluation of simulated hydrograph at four stations of the 2007 event: mean absolute error*

| Gauging station | *MAE* (m) |
|:---:|:---:|
| Dunafoldvar | 0.75 |
| Paks | 1.31 |
| Baja | 1.87 |
| Mohacs | 1.55 |
| Mean | 1.37 |

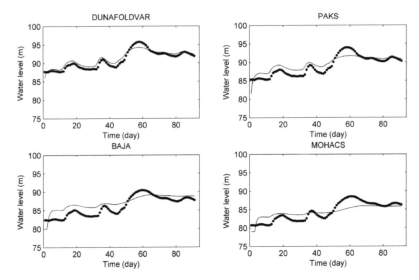

**Figure 4.5** *Simulated water levels hydrograph (solid line, the 2006 event) and in situ water levels (dots) at the four gauging stations using best-fit model (calibrated against radar altimetry) of the 2006 event.*

To better understand the reasons for the poor simulation of the water levels in Mohacs, we also compared the results of the optimal model to the flood water levels observed during the 2006 flood event at the four stream (in situ) gauging stations (Figure 4.5). The results were similar to the ones obtained by simulating the 2007 event (though with lower *MAE*): the flood peak in Mohacs is poorly reproduced and simulated water levels are less variable than the observed ones. At least Figure 4.5 explains that the error in Mohacs is not likely induced by the flood magnitude difference between 2006 and 2007 event, which, highlights the parameter uncertainty when a well-calibrated hydraulic model is used for a second event prediction. A plausible explanation for this effect is the overestimation of floodplain inundation due to the use of a homogenous $D$ along the river reach. This overestimation enhances the attenuation of the flood wave and therefore leads to lower flood peaks downstream and less variability in water levels. Using different, distributed $D$ values along the reach may improve the simulation of observed water levels, but will require arbitrary assumptions, not to mention the issue of potential over-fitting. Lastly, the dykes of Danube River in this area are not likely to be captured by SRTM DEM, so the model allows for accelerated flood water overtopping from main channel to floodplain (which in reality is not happening) – this leads to the lower variability of the simulated water levels in the channel.

## 4.7  CONCLUSIONS

This Chapter tests the ability of a hydraulic model, built and calibrated by using freely available remote sensing data, to simulate flood water levels on the Danube river. In particular, the model utilizes SRTM topography as geometric input, and radar altimetry as calibration data. The performance of the model is found reasonable for large-scale flood studies in data-poor areas, but generally not sufficient for detailed inundation studies, e.g. the design of hydraulic structures. These results contribute to the understanding of the potential usefulness and limitations of radar altimetry data and SRTM topography in supporting hydraulic modelling for water level prediction in data-poor areas.

The study is unavoidably associated with the particular nature and limitations of the case study as well as the assumptions made. Inundation modelling which focuses on flood extent prediction should also be validated against flood extent information. It is therefore recommended to evaluate the model against flood extent maps as well, such as the ones derived from satellite imagery data. In addition, the potential of SRTM topography might not be fully understood without a comparison to high resolution-accuracy topographic data, such as the newly released TerraSAR-X add-on for Digital Elevation Measurement (TanDEM-X) topography. Moreover, future research should corroborate or falsify the results of this study by further exploring the potential and limitations of radar altimetry and global DEMs in supporting flood inundation models across different scales and along diverse flood conditions.

# CHAPTER 5

# SRTM-BASED INUNDATION MODELLING OF A LARGE RIVER

# IN DATA-SCARCE AREAS: REGIONAL VERSUS PHYSICALLY-

# BASED METHODS

Besides the lack of detailed topographic input data, one of the main obstacles in mapping flood hazard in data-scarce areas is the difficulty in estimating the design floods, i.e. peak river discharge corresponding to a given return period. This estimation can be carried out by using two main approaches: (i) data-based regionalization techniques, which are based on time series of flood data collected in regions with similar hydro-climatic conditions, or (ii) physically based models, which aims to simulate the entire cascade of processes from precipitation to runoff. In this chapter we compare these two approaches with reference to a specific case study, a river reach of the Blue Nile between Ethiopia and Sudan where SRTM DEM is used as the topographic input. In particular, we consider two alternative methods: i) regional envelope curve, whereby the design flood (e.g. 1-in-20 flood) is estimated from African envelope curves, and ii) physical model cascade, whereby the design flood is determined from the physical model chain of the European Centre for Medium-Range Weather Forecasts. The comparison is not only made in terms of estimated design flood, but also in terms of resulting flood extents maps produced by hydraulic models based on SRTM DEM. The results show the potential and the limitations of both approaches as well as the challenge of estimating design floods in data-scarce area. We also show the importance of data collection and how even short time series would have a remarkable value in supporting the estimation of design floods.

## 5.1   INTRODUCTION

The planning and implementation of numerous strategies to cope with floods require the estimation of the areas that are potentially flooded. To this end, flood inundation models are used to simulate a design flood, i.e. river discharge corresponding to a given return period. Design floods can be statistically derived from historical records, such as time series of annual maximum flows. However, hydrological measurements are missing in many areas of the world (Stokstad, 1999), and the number of stream gauging stations is declining. Thus, flood hazard mapping and inundation modelling are challenging to be implemented when flood data are not available.

To estimate design floods in ungauged basins (Sivapalan et al., 2003), two main categories of approaches can be identified: i) more traditional approaches, such as regionalization methods, which are based on the elaboration of flood data collected in regions with similar hydro-climatic conditions, and (ii) emerging approaches, such as the use of large-scale physically based models, which aim to simulate the entire cascade of processes from precipitation to runoff.

### 5.1.1   Regional envelope curve

An example of the first category is the empirical regional envelope curve (hereafter REC), which is a traditional method to estimate extreme floods in ungauged basins (Castellarin, 2007). REC is based on the concept of transferring the hydrological knowledge from gauged catchments to ungauged (or poorly gauged) catchments and provides an effective summary of regional flood experience (Castellarin, 2007; Di Baldassarre et al., 2006). The estimated design floods (inflows) can then be used to feed a hydraulic model and generate flood hazard maps in data-scarce areas (Brandimarte et al., 2009; Padi et al., 2011).

### 5.1.2   Physical model cascade

An example of the second category is the use of physical model cascades (hereafter PMC). Pappenberger et al., (2012) recently used a large scale PMC and derived, at the global scale, river discharge values corresponding to different return periods. This method included the derivation of meteorological forcing data and discharge acquisition from a land surface model based on a 30 year (1979-2010) simulation period.

This study aims to compare the capability of REC and PMC in estimating design floods to map flood hazard in data-scarce areas. The design floods estimated from the two methods are benchmarked with that estimated from in-situ data. The comparison between the two approaches is not only conducted in terms of the estimated design floods, but also in terms of the resulting flood extents maps. To this end, the comparison is made by using a river reach of the Blue Nile as a case study.

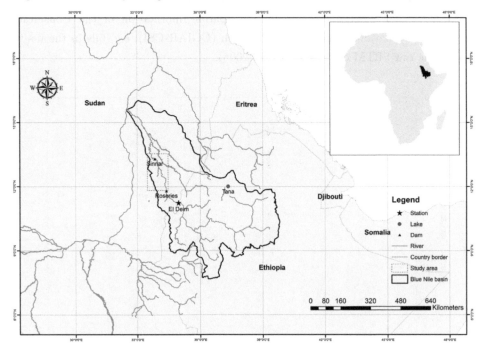

*Figure 5.1* *The Blue Nile basin and study area*

## 5.2 STUDY SITE AND DATA AVAILABILITY

This study is carried out on a river reach of the Blue Nile around 280 km between Rosaries Dam and Sinnar Dam in Sudan, with a mild slope of about $0.12 \times 10^{-3}$ (Figure 5.1). The historical discharge data is available at gauge station El Deim near the Sudan-Ethiopia borders, which is about 70 km upstream of Rosaries Dam. The rainfall in the Blue Nile basin is confined in the single season, consequently the river flows are concentrated in a short period (Sutcliffe and Parks, 1999). The available in-situ data of river discharge, provided by the Ministry of Water and Energy of Ethiopia, is a time series of 25 years of annual maximum discharge covering the whole rainy season (from June to September) (Baratti et al., 2012). This river station is treated as if it was ungauged in the application of REC and PMC, while design floods estimated by in-situ data are used as a reference.

The topographic data of the test site are derived from the global SRTM Digital Elevation Model (DEM). Although SRTM has a coarse resolution, a number of studies showed its value for large scale flood studies (e.g. Sanders, 2007; Neal et al., 2012a; Yan et al., 2013). The DEM used in this study as input to the hydraulic model is post-processed by the Consortium for Spatial Information of the Consultative Group for International Agricultural Research (CGIAR-CSI), e.g. fills in the no-data gaps in the raw SRTM data (Javis et al., 2008).

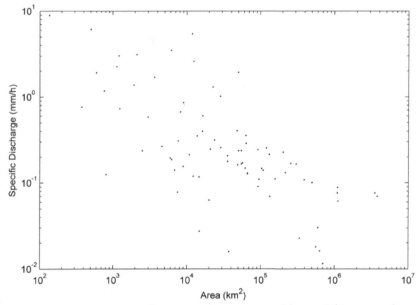

**Figure 5.2** *Flood experience of 1-in-20 year event in Africa: Q/A in mm/h and A in km²*

## 5.3   DESIGN FLOOD ESTIMATION

### 5.3.1   Design flood derived from REC

Padi et al., (2011) analyzed African flood data and derived regional envelope curves for the African continent. The flood experience in the African continent (1-in-20 year, Figure 2) is described by Equation 5.1:

$$\ln\left(\frac{Q}{A}\right) = a + b\ln(A) \qquad\qquad (5.1)$$

where Q is the discharge value, A is the drainage area, while a and b are the intercept and slope of the line. It should be acknowledged that the African catchments used to derive the REC (Figure 5.2) do have a variety of hydro-climatic conditions. Thus, treating them as a homogeneous region can be questionable. Merz and Bloschl (2008a, b) and Viglione et al., (2007), for instance, have proposed the concept of flood frequency hydrology, highlighting the relevance of combining flood data with causal information on flood processes. However, as the goal of this study is to compare a purely data-based approach with a physically-based one, physical reasoning is not introduced in the regional approach.

### 5.3.2   Design flood derived from PMC

In PMC, the European Centre for Medium Range Weather Forecasts (ECMWF) land surface model, the Hydrology Tiled ECMWF Scheme of Surface Exchanges over Land (HTESSEL, Balsamo et al., 2012) is coupled with ERA-Interim reanalysis meteorological forcing data to generate runoff for the 30 year (1979-2010) simulation period. ERA-Interim/Land is the result of applying the land-surface model simulation using HTESSEL, with meteorological forcing from ERA-Interim and precipitation adjustments based on the Global Precipitation Climatology Project (GPCP) v2.1 product (Huffman et al., 2009). ERA-Interim/Land preserves closure of the water balance and is therefore more suitable for climate applications than the land surface parameters included in the original ERA-Interim data set (Balsamo et al., 2012). River discharges are produced for the 30 year time period (1979-2010) using CAMA-Flood (Yamazaki et al., 2011), a global river routing algorithm. CAMA-Flood simulates horizontal water transport by using a diffusive wave equation and backwater effects are also taken into account. The discharges are calculated for a 25 km grid resolution and downscaled via catchment based rating curves to 1 km resolutions. A Gumbel (EV1) distribution function is fitted to the annual maximum discharge of 30 years to derive the design floods of the study area (Pappenberger et al., 2012).

## 5.4   HYDRAULIC MODELLING

To assess how errors in the estimation of design flood propagate into flood hazard mapping, the estimated design flood is used as hydrological input of a hydraulic model. In particular, we use the LISFLOOD-FP (Bates et al., 2010) two-dimensional (2D) hydraulic code, which solves simplified version of the shallow water equations that preserve acceleration, but neglects advection term. LISFLOOD-FP is utilized to propagate the 1-in-20 year flood discharges derived by two methods (i.e. REC, PMC) as well as the one estimated from stream-gauge data. Given the absence of major tributaries and large storage areas, it is assumed that the discharges measured at El Deim station are representative of the ones entering the Rosaries reservoir. Also, simulations are run by assuming steady flow. This is justified by the long duration of flood events in this river reach. Also, it allows avoiding hypotheses on the shape of the design hydrograph. The simulation periods are set as long enough to ensure that maximum inundation extents are obtained for all simulations. A water surface slope is used as downstream boundary condition.

As SRTM's Interferometric Synthetic Aperture Radar (IfSAR) technology cannot detect channel geometry beneath water surface, the bed elevation of main channel is not properly captured by the DEM (Farr et al., 2007). This can be compensated by parameters (i.e. Manning's roughness coefficients) in hydraulic modelling (Petersen and Fohrer 2010, Yan et al., 2013, Pappenberger et al., 2006, Schumann et al., 2007). However, the scope of this modeling exercise is not to accurately simulate inundation patterns, but rather to get insights into the magnitude of the discrepancies in terms of flood extent when using different estimations of the design flood. Thus, also in view of the absence of flood extent observations, the Manning's roughness coefficients are set as uniform, with standard values of 0.04 for the main channel and 0.06 for the floodplain (Chow, 1959). Hence, all simulations are based on the same parameters, initial and boundary conditions, while they different only in terms of inflow, equal to the estimated design flood.

A simple aggregate performance measure is used to compare the flood extents simulated using the design flood estimated by REC and PMC to the reference ones obtained with in-situ stream gauge data (see Equation 1.5).

## 5.5 RESULTS AND DISCUSSION

The 1-in-20 year design flood determined from in-situ stream gauge observations of annual maximum flows using a Gumbel distribution is around 10,600 $m^3s^{-1}$. It should be noted that this value is affected by uncertainty. Although the length of the time series allows a robust estimation of the 1-in-20 year flood, multiple sources of uncertainty affect the flood data used for such estimation, such as river discharge measurement errors caused by extrapolation of the rating curve and changes of the cross-section due to sediment transport (ElMonshid et al., 1997). The quantification of this uncertainty is not an easy task and out of the scope of this study, but the literature suggests that errors around 20% can be expected (Di Baldassarre and

Montanari, 2009). Thus, a range of $10,600 \pm 2,120$ m$^3$s$^{-1}$ is used here as a reference for the 1-in-20 year flood.

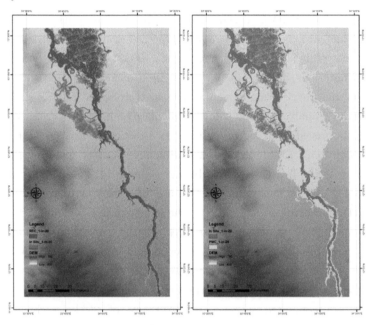

***Figure 5.3*** *Comparison of In-situ flood extents of 1-in-20 year return period to REC (left panel) and PMC (right panel)*

The design floods estimated by REC and PCM differ markedly from this reference values. The application of the REC leads to an estimate of 5,500 m$^3$s$^{-1}$ and therefore largely underestimates the reference value; while the PMC leads to an estimate of 24,100 m$^3$s$^{-1}$ and therefore largely overestimates the reference value.

Figure 5.3 shows the impact of these differences in design floods in terms of the resulting flood extent. The REC-based inundation model underestimates the reference flood extent ($F$ equal to 0.56). While the PMC-based model leads to overestimation ($F$ equal to 0.39).

The underestimation of design floods associated to the application of REC might be the consequence of the inherent space-time heterogeneity of ungauged basins. Even though the hydrological information come from the two large databases (UNESCO 1984; IAHS 2003) are believed to be reliable (Padi et al., 2011), the catchments are not entirely homogeneous in terms of hydro-climatic conditions. Thus, the empirical envelope curve might miss the hydro-climatic characteristics of the Blue Nile catchment which is dominated by monsoon climate in the Ethiopian Highlands. REC, which is essentially a logarithmic linear regression (equation 1), might induce systematic error in design floods estimation.

The overestimation of PMC design floods is likely due to the large uncertainty accumulated and propagated through the process-based meteorological-hydrological model chain. As stated by Pappenberger et al., (2012), the design floods derived by PMC have limitations due to the fact that it is affected by uncertainties in each component of the physical model chain. The meteorological forcing time series (i.e. the ERAInterim reanalysis) is too short to calculate high return periods and might be substantially improved by using an enhanced correction routine or better correction data. In addition, ERAInterim has difficulties with representing accurate precipitation patterns in this region (Mwangi et al, 2014). The quality of the input data set could also be improved by the use of downscaling or other correction techniques (Di Guiseppe et al., 2013). The land surface scheme (i.e. HTESSEL) results are hampered by the uncertainties in the aspects of model structure, parameters, grid resolution, numerical scheme and topographic data. The model structure is never perfect in hydrological modelling and the large number of parameters lead to considerable uncertainties as well as equifinalities. The surface water fluxes are computed in a coarse grid size of 25 km, which adds more uncertainties to the upstream boundary condition in a 90 m-resolution hydraulic model of LISFLOOD-FP. All those uncertainties contribute to largely overestimated design floods and flood extents compared to the ones derived from in-situ data (Figure 5.3).

This study shows that estimating flood hazard in data-scarce areas is a real challenge, as both methods used to estimate the design flood for a reach of the Blue Nile have demonstrated clear limitations. In his invited talk at the 2014 EGU General Assembly, Demetris Koutsoyiannis made a provocative point that the best way to make predictions in ungauged basins is to make them gauged (Koutsoyiannis, 2014). However, one may argue that, to properly estimate design floods, long time series are needed. Thus, we made an additional experiment to address the following research question: if a stream gauge was installed, how long would it take to get design flood estimations more reliable than the ones obtained using either REC or PMC?

To explore this point, we randomly sample n discharge values from the entire time series of annual maximum flows. The 1-in-20 year design flood is then estimated based on the chosen n available observations by fitting the Gumbel distribution and using the method of moments. For each $n$ observations, we repeat the random sampling a large number of times (i.e. 10,000): The estimations are summarized in terms of 10th, 90th percentile, as well as the median, in Figure 5.4. Note the uncertainties due to flood data, the choice of extreme value distribution and other sources are out of the scope and not considered in this study.

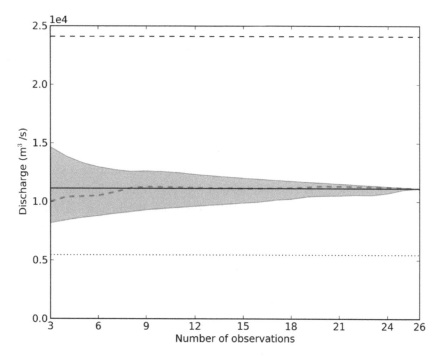

**Figure 5.4** *1-in-20 year design flood estimation depending on the number of observations (ranging from 3 to 26). The black solid line is the value estimated using 26 annual maximum discharges (uncertainties due to flood data and other sources are neglected and not shown here); the grey area is the uncertain estimation bounded by 10th and 90th percentile; grey dash line is the median; black dash line is the design flood estimation using PMC, black dotted line is the design flood estimated using REC.*

In Figure 5.4, the uncertainty bounds reduce significantly with the increasing number of observations, particularly when n is equal or below 6 (with first 3 added observations result in about 50% of the uncertainty bounds reduced, compare to the estimation using 26 observations). The reduce rate of uncertainty bounds decreases substantially when n is equal or above 10 (e.g. 21% and 17% of the uncertainty bounds are reduced for 10 to 18 and 18 to 26 observations considered). This indicates that the rate of increasing information content is larger when only a few observations are available, which highlights the value of newly collected data with limited initial available observation in data-scarce areas. The most interesting outcome is that the estimated design floods using limited observations are better than the ones derived from either REC or PMC, even with only three observations (Figure 5.4). Similar results were found in Brath et al., (2009), in a rather different context, where the direct method to estimate index flood with very limited information over-performed statistical models and physically-based models. This also highlights the importance of data collection for design flood estimation. Given the rapid development of data collection technology (e.g. floating acoustic Doppler velocity measurement devices), this study suggests that the efforts to make more rivers gauged are at least as useful as efforts to improve the methods for runoff estimation in ungauged basins.

## 5.6   CONCLUSIONS

The aim of this study is to investigate the potential and limitations of a traditional method (i.e. REC) and an alternative method (i.e. PMC) for estimating design floods in a river reach of the Blue Nile. To get insights on the impact on flood hazard studies, the discrepancies in design flood estimations are also evaluated through the comparison of the resulting flood extent.

The results of this study show that PMC model overestimates the design flood extents compared to the benchmark in-situ model, while REC underestimates the design flood extents. Given the over- and under- estimation of flood extents, in this particular case study, the results indicate that flood extents derived by PMC and REC might not be appropriate to support flood risk management (and detailed land use planning) at the local scale. While robust conclusions about the appropriateness of those methods require additional cases studies, these outcomes underline the challenge of assessing and mapping flood hazard in data-poor areas. Moreover, we show the importance of collecting data as we demonstrate that in this test site even very short time series can provide a better estimation of the design flood than the one provided by other methods.

# CHAPTER 6

## SYNTHESIS, CONCLUSIONS AND FUTURE RESEARCH

This thesis attempts to explore the potential and limitations of low-cost space-borne data in supporting hydraulic modelling of floods. As data and other model aspects (e.g. uncertainty, modelling purpose) are deeply intertwined, some key issues related to these data in hydraulic modelling has to be understood before being integrated in the flood hydraulic modelling practice. For example, the DEM resolution might be less of an issue affecting model results compared to its vertical accuracy. Other sources of uncertainties explicitly considered in hydraulic modelling would interact with topographic uncertainty and affect overall model performance. Moreover, model results based on simple model and coarse data might not be worse than that based on complex model and fine data. In the end, modelling purpose dominates the choice of data and modelling tools. In this chapter we summarize these issues as potential and limitations together with conclusions and future research in this chapter.

## 6.1 SYNTHESIS

New data sources are becoming increasingly important to support flood studies in data-scarce areas. At the same time, new technologies in remote sensing are developing at a fast pace towards ever-increasing accuracy, resolution, frequency and low cost. For example, the upcoming topographic data set from TanDEM-X (with a spatial resolution of 12 m and relative vertical errors less than 2 m, and at fairly low cost) may prove invaluable for scientists and practitioners with an interest in flood mapping and modelling. This thesis attempted to explore the usefulness of low-resolution, low-cost space-borne data in supporting flood inundation modelling.

The limitations and potential of data should be discussed in view of various model aspects. On the one hand, for instance, SRTM`s global coverage, zero costs, easy acquisition, are obvious reasons that make this type of data attractive. On the other hand, the actual value of this data depends on the scale and purpose of the model, and other sources of uncertainty affecting the modelling exercise. This condition also applies to satellite imagery and radar altimetry, which have been investigated in this thesis.

### 6.1.1 DEM resolution and accuracy

Global coverage DEMs with low vertical accuracies (Table 1.2) are typically not useful for local detailed floodplain inundation studies requiring accurate and precise estimations of high water levels (e.g. modelling supporting the design of a new bridge). However, as SRTM DEM at 90 m resolution, for instance, has been shown to be as accurate as 2 m errors in the vertical in some floodplain areas (Schumann et al., 2013), its use on large scale flood inundation studies can be promising. This thesis shows that this is actually the case, especially if other major sources of uncertainty are explicitly considered (Chapters 2 and 3). Also important to note is that remote sensing acquisition of topographic height data means that undesirable features related to the data processing (voids and artificial hills) or related to the data acquisition process (e.g. vegetation canopy) need to be removed or corrected. For LiDAR processing such techniques are well advanced and lead to very accurate bare ground DTM; however in the case of SRTM for example, vegetation removal procedures are much less understood and only in a research phase meaning that the SRTM DEM is currently not available as a bare ground DTM by default and vegetation needs to be removed before the DEM can be applied to flood inundation mapping or modelling (Baugh et al., 2013).

Despite the limitations noted above, the resolution of the topographic data is typically less significant than accuracy for providing reasonable modelling results for large-scale flood studies (e.g. Horritt and Bates, 2001), as long as the resolution of the DEM is not too coarse to capture the main channel of the river (e.g. Sanders 2007) and a global performance measure (e.g. Aronica et al., 2002) is used. It should be mentioned that the recent advances in hydrodynamic modelling bring improvements to models that can credibly simulate flood processes at sub-grid scale using coarse resolution and lower accuracy DEMs (e.g. LISFLOOD-FP sub-grid channel (SGC) as developed by Neal et al., 2012). This opens up new opportunities in the exploitation of coarse resolution data.

Performance measures that take into account local weights which vary among locations according to their importance often allow for better estimation of flood hazard, but require more detailed topographic data. For example, the optimum calibration for flood extent was found to be quasi invariant with respect to changes in DEM resolution in River Severn, UK. In particular, the model reached maximum performance at a resolution of 100 m, after which no improvement can be achieved with increasing resolution (Horritt and Bates, 2001). For the River Dee in Wales (UK), the predicted flood extent by using LiDAR data with resolution of 20 m is very similar to that obtained by aggregating the DEM to 80 m (Chapter 2). Although Stephens et al., (2013) argued that a global binary pattern measure for flood extent evaluation might be biased depending on the size of the flood and shape of the catchment, DEM resolution is often less of an issue than its vertical accuracy. Moreover, given low-cost SAR images used for hydraulic model calibration and validation are often of coarse resolution, the DEM used for building the model should be of a similar (low) resolution. In other words, a fine resolution DEM-based model (e.g. built from LiDAR) would lose its advantage in terms of resolution when low resolution imagery is used for model calibration. Ideally, the spatial resolution of low-cost space-borne images and DEMs should develop with similar pace, to be fully integrated in hydraulic modelling. The case studies presented in this thesis are representative of many rivers and floodplains in Europe.

### 6.1.2   Treatment of other sources of uncertainty

The uncertainty introduced by inaccurate topographic data can be as significant as the other sources of errors in river flood studies, including model parameters (e.g. roughness coefficients, Beven and Binley, 1992; Beven and Freer 2001; Montanari 2007), hydrological input (e.g. river discharge values, Di Baldassarre and Montanari, 2009) and failure of flood defences (e.g. levees or dikes, Di Baldassarre et al., 2009c;

Mazzoleni et al., 2014). However, it is still unclear how topographic uncertainty interacts with other major sources of uncertainties. Given a complicated, non-linear system, several sources of uncertainty might strengthen or compensate each other, which result in amplified or narrowed uncertainty bounds. There are very few studies that analysed at this issue (e.g. Yan et al., 2013 (Chapter 3); Jung and Merwade, 2012), which is therefore worth exploring further. Topographic uncertainty might be compensated by other sources of uncertainties in hydraulic modelling if they are explicitly taken into account (Chapter 3).

### 6.1.3 Equifinality and data-model relation

The concept of equifinality in flood inundation modelling is that models with different parameter sets (usually roughness coefficients in channel and floodplain) are equally able to simulate flood extent. As SAR imagery provide less detailed information than the high resolution ones, it is more likely to fit the less informative flood extents equally well with models based on more parameter sets. In other words, more information is needed to differentiate models with different parameter sets. Thus, equifinality could get more severe when flood inundation models are conditioned on low resolution SAR imagery. However, on the other hand, when high resolution SAR imagery (detailed information) is available, but they are beyond what several models (with their structure and parameters) can mimic, model with different parameter sets values could behave 'equally bad'. For example, models cannot simulate detailed inundation patterns presented in SAR regardless the model parameter spaces and the model parameter sampling techniques. Fabio et al., (2010) use 390 flood depths points to calibrate a 2D flood propagation model. The equifinality of the parameter sets showed that the model structure is not sufficiently complex to properly describe the information content of the calibration data. In this context, coarse resolution data does not look really bad in hydraulic modelling.

The exploration of information content to effectively improve hydraulic modelling is a topic of great interest during recent years (Dottori et al., 2013). More detailed information does not necessarily lead to improved model performance. Besides, Neal et al., (2012b) demonstrated that in some circumstances simpler models might achieve accuracies similar to that of more complex model. In a model conditioning and evaluation context, simpler models with coarse-resolution low-accuracy topographic data might even over perform that based on high-resolution high-accuracy ones (Mukolwe et al., 2014). Those examples illustrate that simply bringing in more detailed data does not necessarily solves neither the issue of equifinality, nor improves the model performance. Thus, a proper combination of model structure, parameter and input/calibration data is a better direction for flood inundation modelling. Despite the recent progress on satellite constellation (the potential of Sentinel-1 and COSMO-SkyMed are yet to be explored), for the time being there is still little chance to obtain more than one low-cost flood event map per event. Therefore, with such calibration data, neglecting some physical processes that are less important (e.g. secondary circulation, advection term etc.) seems to be a viable option (e.g. Bates 2012). In this context, low-cost data in combination with the right modelling tools producing reasonable good results would be the preferred case for flood studies in data-sparse areas (Chapter 2, 3, 4). Essentially, this is also an issue of how input (data) uncertainty interacts with model structure uncertainty.

### 6.1.4　Flood frequency and micro-topography

Hydraulic modelling based on SRTM topographic data can provide reasonable results for large-scale flood studies. This thesis (Chapter 2, 3, 4) and recent studies have explored the potential of this global and freely available topographic dataset. However, the term "reasonable" requires a number of conditions. First, the frequency of flood events is important. Most studies reviewed herein focused on the simulation of low-frequency floods (e.g. 1-in-100 year), which most likely do not require high-resolution and high-accuracy DEMs for detailed descriptions of small topographic features, which drives the inundation processes in case of high-frequency floods (e.g. 1-in-2 years). This is due to the fact that e.g. low-frequency floods can submerge some small topographic features, by which flood patterns would be less affected due to large accumulated flood volume on floodplain. For a more comprehensive analysis of the utility of SRTM DEM in hydraulic modelling, there is thus still a need to explore the higher-frequency events even though they are less commonly studied due to their more limited impact on society. Such a thorough analysis may also contribute to the understanding of the limitations of low accuracy topographic data in support of hydraulic modelling. On the other hand, by understanding this limitation, SRTM might offer great potential to large-scale, low-frequency flood modelling (e.g. Chapter 3, 4).

### 6.1.5　Absence of in-channel data

As channel conveyance has a significant impact on modelling flood processes and SRTM (or even LiDAR for that matter) cannot estimate channel bed elevation, there is the need to approximate channel bed prior to hydraulic modelling (e.g. Neal et al., 2012a; Patro et al., 2009; Yan et al., 2015a). The first global river depth and width data set (Andreadis et al., 2013; Schumann et al., 2013) could significantly contribute to main channel information. However, the large uncertainties of this dataset (see earlier) would limit its utility in hydraulic modelling studies. The hydraulic geometry theory, which describes a power law relationship between channel width and depth, is a promising alternative to estimate main channel information (e.g. Neal et al., 2012a). However, it is also the source of major inaccuracy in water level simulation, particularly at low flow conditions. In this thesis (Chapter 4), we propose an approach to lower the channel bed elevation for $D$ meter and treat $D$ as a model parameter. This allows an improved simulation of water level in terms of $MAE$ in model evaluation. Although the optimum model based on this approach cannot capture the flood peaks in the downstream station, it is still a relevant attempt to improve in-channel part of SRTM by parameterizing the overestimation. Nevertheless, there is still the need to come up with more reliable approaches to estimate channel geometry in the absence of field surveys.

### 6.1.6 Modelling purpose matters

SRTM-based hydraulic models can reach reasonable levels of accuracy in terms of water level and flood extent prediction, however, these are still of much lower accuracy than high-resolution and high-accuracy DEM-based models. There is still no broad consensus on the reliability and accuracy criteria for hydraulic models (Hunter et al., 2007). In fact, the study aim should dominate the requirements of modelling accuracy. Modelling purposes is deeply intertwined with appropriate topographic data and flood extent data. Coarse resolution model can hardly be useful in urban flood modelling where detailed description of flood hazard at local scale is required (e.g. Neal et al., 2009). Urban areas are characterized by critical length scales which are determined by both the shortest building dimension and the distance between buildings. Model performance deteriorates significantly at resolutions above these thresholds (Fewtrell et al., 2008). Thus, micro-topography which cannot be represented by global DEMs is vital for controlling flood pattern, while its impact becomes less relevant in large-scale flood studies. Meanwhile, high resolution SAR imagery is also essential to constrain the uncertainty in high resolution flood models.

Therefore, modellers should carefully decide whether SRTM DEM (and other space-borne data) is appropriate for the study at hand according to their modelling purposes. For instance, the design of hydraulic structures often requires detailed modelling of a river reach. In this case, SRTM data are most likely not suitable. Yet, SRTM data can be useful to support hydraulic modelling in data-scarce areas when the purpose is to generate large scale flood maps as a non-structural measure to raise flood awareness in flood prone areas. For example, the devastating India-Pakistan flood in 2014 has killed 277 people and caused huge economic losses in Jammu and Kashmir, where almost no flood data are available whatsoever. The results provided by a large-scale flood model using space-borne low-cost data (e.g. with the water level accuracy 1-1.5 m) can at least build/raise citizens' flood awareness towards preventing potential losses on such huge event, or quantify the after-event flood damage.

The issue of data scarcity for flood modelling is not only economic and technological, but could also be highly political in a geopolitical sensitive area like Kashmir. In this context, the potential of space-borne, low-cost data which is freely and globally accessible online is enormous. Similar situations could be found in the Nile basin where many topographical and hydrological data are kept confidential and not shared across countries. This makes integrated trans-boundary flood risk management very difficult. The emerging low-cost space-borne data with their demonstrated potential in large scale river studies can be one of sources for overcoming these lacks of trans-national data.

Yet another potential application of technologies covered in this thesis is insurance industry. Flood insurance focuses on the importance of flood prediction in the context of the business models of insurance companies. They need to manage large quantities of capital reserves, which have to be paid out in case a flood disaster strikes. This requires insurers to classify areas with a variety of land uses to several levels of risks and throughout a range of countries, perhaps even in the continental scale. As discussed in this thesis, SRTM seems to be currently a reliable global DEM that could be utilized for flood hazard estimation (TanDEM-X is still under development when this thesis is drafting), which is an essential task in Flood risk mapping for (re)insurance companies.

There is currently no guideline for matching level of model accuracy with specific modelling needs, not to mention those guidelines would perhaps be different when applied by people with different visions of various institutions in different countries. However, the conclusions currently drawn from SRTM-based hydraulic modelling studies could provide useful reference and guidance, and contribute to future studies in data-scarce areas.

## 6.2 CONCLUSIONS IN BRIEF

### 6.2.1 Main conclusions

Low-cost space-borne data (e.g. SRTM topography, SAR imagery, radar altimetry) provides a promising potential for hydraulic modelling of floods in data-scarce areas. In particular, the performance of flood inundation models based on SRTM topographic data can be close to the one of models based on high-resolution high-accuracy DEM (Chapters 2, 3 and 4). However, their actual usefulness could be affected by several factors, such as the scale of the river under study, the specific modelling purpose, flood frequency, availability of in-channel information, other sources of uncertainty considered, as well as the choice of modelling tools. Thus, estimating flood hazard in data-scarce areas is still a real challenge (Chapter 5). There is no general consensus about the level of model accuracy that should be achieved for any specific modelling purpose, not to mention this would probably differ from country to country by people with various visions. Therefore, users should carefully consider those factors discussed in this thesis before using low-cost space-borne data in hydraulic modelling of floods.

### 6.2.2 Specific findings

Besides the main conclusions, a number of findings of this thesis are listed below:

- DEM resolution is often less important than its vertical accuracy for hydraulic modelling in large-scale flood studies, as long as the key topographic features (e.g. embankments) are represented (Chapter 2).

- The prediction of 1-in-200 year flood profiles in the study area of River Po, using the HEC-RAS model based on SRTM topography data, are within the accuracy that is typically associated with large-scale flood studies, as long as the other sources of uncertainty are explicitly considered (Chapter 3).

- The performance in reproducing in-situ water levels for large-scale flood studies in data-scarce areas of a hydraulic model built and calibrated by using SRTM and radar altimetry was found reasonable, but generally not sufficient for detailed flood studies (Chapter 4).

- Flood extent maps derived by PMC and REC might not be appropriate for flood risk management (and detailed land use planning) in ungauged basins. As methods based on ground data with a very few observations over performs PMC and REC, the importance of data collection is highlighted (Chapter 5).

## 6.3   RECOMMENDATIONS

This thesis explores the potential and limitations of low-cost space-borne data for inundation modelling of floods under uncertainty. Model performances based on those data are quantified on three cases covering from medium to large scale of rivers. It is still hard to draw solid conclusions based on limited number of studies. Thus, more evidences should be collected to support the general findings. Moreover, the scale here mentioned does not only refer to spatial scales, but should also include scale of floods (i.e. flood frequency). The more case studies we have, the more detailed pictures showing potential of low-cost space-borne data supporting flood modelling could be drawn.

Besides the two dimensions (i.e. spatial scale and flood frequency), the third important dimension (i.e. modelling purpose) should also be added to the big picture. In this thesis, the usefulness of space-borne data is qualitatively discussed according to various modelling purposes, while it is also worth to be quantified. It is therefore recommended to conduct more similar studies but with end-users' (stakeholders') feedback on specific performance requirement and satisfaction of existing model accuracy. The results should be rescaled from 0 to 1 among studies with various purposes. A 3D figure with river scale, flood frequency and modelling purpose could therefore be plotted.

The current efforts focus on quantifying the potential of low-cost space-borne data, measures to improve model performance based on existing data should also be explored. In particular, as the upcoming satellite missions (e.g. TanDEM-X) that provide more accurate global topographic data are still unlikely to include in-channel geometry information, more research should dedicate on resolving this issue of SAR-based DEMs support flood modelling. The current solutions are often associated with large uncertainty, therefore improvement is needed.

Lastly, flood modelling and remote sensing is indeed a hot topic in hydrology and provides huge potential to shape the way we model floods. It seems that current researches focusing on remote sensing data are unavoidably falling into a 'data-emerging, data-testing' circle. It is indeed worth to explore the new data, while on the other hand, lack of novelty. Therefore, an innovative methodology that could discover the essence of 'remote sensing data - model relation' is badly needed.

## 6.4 DATA FROM THE FUTURE SATELLITE MISSIONS

As the space-borne data is moving with a rapid pace towards high-resolution, high-accuracy and low-cost, more and more researches are focusing on the integration of those fast-growing data into inundation modelling. In the light of advances of remote sensing technology in hydrology, we believe that numerical modelling of floods has entered a whole new stage. The section below introduces the upcoming future satellite missions which could likely transform remote sensing hydrology.

### 6.4.1 TanDEM-X

TerraSAR-X add-on for Digital Elevation Measurement (TanDEM-X) is the new world-wide DEM of DLR (German Aerospace Center) that provides homogeneous pole-to-pole coverage of unprecedented accuracy: at a spatial resolution of 10-12 m and a relative vertical accuracy of less than 2 m on slopes less than 20%, and 4 m on slopes greater than 40% (Eineder et al., 2012, Figure 6.1). These specifications exceed any other global satellite-based elevation model available today, compared to 30/90 m spatial resolution of SRTM. Up to now, the DEMs of the most of Australia, North America and Russia have been finished. Areas in South Africa and South America are next in the production. It is expected to complete the global DEM by the end of 2015 (Zink 2014). Even though the release of this DEM is currently under a commercial license which is operated by Airbus Defense & Space, the associated low cost (100 euro per quota, ~7700 km$^2$) would be a relatively minor obstacle if cost were a constraint. Scientific assessment of this DEM is at the moment only in an experimental phase, but there are already several assessments of the Intermediate DEM (IDEM), which is the first calibrated and mosaiced product of TanDEM-X. Table 6 summarizes the IDEM quality assessment of TanDEM-X that covers areas with various characteristics. The results show that in hilly and sparsely vegetated areas the accuracy requirements are already fulfilled, whereas for flat and sparsely vegetated areas they are even better. In hilly and vegetated regions the accuracies can almost be achieved (Gruber et al., 2012). This is only the intermediate product of TanDEM-X based on one coverage, the larger phase errors in hilly and vegetated areas can be corrected in the upcoming second coverage. In addition, TanDEM-X uses Polarimetric SAR interferometry (Papathanassiou and Cloude, 2001) that combines interferometric with polarimetric measurements to gain additional information from semi-transparent volume scatterers. This technique allows extraction of vegetation density and vegetation height. TanDEM-X will be the first mission to demonstrate this technique in a single-pass data acquisition mode (Krieger et al., 2009). Therefore, first observations seem to indicate that TanDEM-X is a promising low-cost alternative and might allow for the first time more detailed local flood studies at the global scale, such as urban flood modelling, flood reinsurance in urban areas, flood study with the purpose of hydraulic structure design etc.

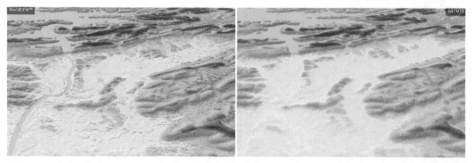

**Figure 6.1** *Comparison of TanDEM-X (left panel) and SRTM 30m (right panel) topographic data in Arkansas, USA (figure taken from WorldDEM image gallery)*

### 6.4.2   Sentinel-1

Sentinel-1, launched in April 2014, is the first of satellites series from the Global Monitoring for Environment and Security (GMES) program, providing operational monitoring information for environmental and security applications (Figure 6.2). Sentinel-1 has been designed to address primarily medium- to high-resolution applications through a main mode of operation that features both a wide swath (250 km) and high geometric (5×20 m) and radiometric resolution. Sentinel-1's revisit frequency and coverage are dramatically better than those of the ERS-1/2 SAR, and the ENVISAT ASAR. The two-satellite constellation offers six-day exact repeat and daily full coverage from north of +45°N to south of -45°S. Its effective revisit and coverage performance may be further improved by access to the planned Canadian C-band SAR constellation. Quick data delivery allows near-real time data to be delivered within 3 h of observation, as well as emergency operations to be completed typically within 24 h. Most importantly, Sentinel-1`s open and free data policy would be extremely beneficial for scientific users across the world, particularly for developing countries. Sentinel-1 operations shall fill the gaps left by the finished operational lifetimes of ERS/ENVISAT and last for a period of at least 7 years (Torres et al., 2012).

Sentinel-1 opens up many opportunities to improve flood mitigation, flood forecasting and modelling, damage evaluation and risk assessment. It could be directly integrated into existing disaster management systems for flood monitoring. Working together with the upcoming TanDEM-X topography as input and calibration data, Sentinel-1 with its free access data policy might allow the first time detailed urban flood modelling in a global scale.

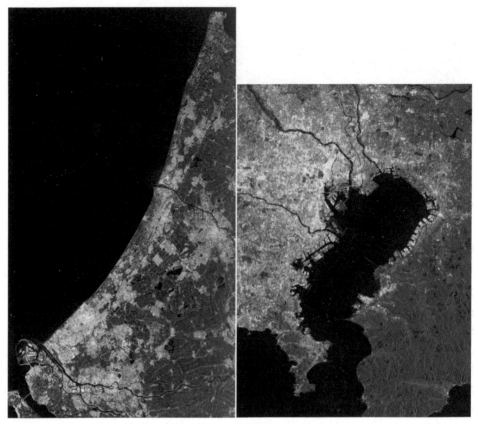

***Figure 6.2*** *Santinel-1 radar imagery show west of the Netherlands (left panel) and Tokyo bay (right panel)*

### 6.4.3 SWOT

The upcoming SWOT mission (expected in 2019) will focus on better understanding of oceans and terrestrial surface water, as well as how water bodies change over time at global scale. The SWOT mission will produce $\pm50$ cm height accuracies per pixel, and allow sampling of rivers at least at 100 m in channel width and perhaps smaller (Bates et al., 2013). The SWOT mission is expected to penetrate vegetation through canopy openings and will also make it possible to estimate discharge in rivers more accurately. With the existing satellite missions, floodplain water levels and its dynamics (rising and falling of water) during the flood event are rarely measured during the passage of a flood wave. SWOT aims to provide water level records for any flooding events that underlay a given satellite overpass. The proposed SWOT satellite mission would have the potential to address many scientific questions and perhaps transform our understanding of hydrology (Bates et al., 2013).

## 6.4  FUTURE RESEARCH

Besides the general findings of this thesis, which provide useful information for the use of low-cost space-borne data, some possible research topics are yet to be explored. This section summarizes a number of potential avenues in this field.

*The impact of flood frequency on the accuracy of inundation modelling based on SRTM topographic data*

Given SRTM`s coarse spatial resolution and poor vertical accuracy, SRTM likely misrepresent some small but key topographic features which could dominate the flood inundation pattern on floodplain. How significant those topographic features are for simulating flood pattern seems to be associated with the flood magnitude (or flood frequency) of the event been studied. Given low-frequency flood waters usually submerge microtopography, by which flood patterns would be less affected due to large accumulated flood volume on floodplain. That is to say, topographic features misrepresented by SRTM that are submerged by a low-frequency flood might become the key obstacles confining a high-frequency flood. Therefore, SRTM might be good at simulating large flood event but not the small ones which 'amplify' the consequences of the lack of small topographic features. The future study tries to explore the interplay between flood frequency and performance of inundation modelling by using SRTM as topography input in two case studies: River Po in Italy and River Dee in UK. A high resolution, high accuracy LiDAR topography-based model is used as the benchmark in comparison with SRTM-based model. Floods with several return periods are tested and a global performance measure is used for evaluating the inundation. The outcome is expected to associate flood frequency of the event under study to the level of accuracy, when using SRTM as the topographic input for inundation modelling.

*Vegetation removed bare earth DTM in flood inundation modelling*

Vegetation canopy, as an undesirable feature for a bare earth DTM, needs to be removed before integrated in flood inundation models. However, current understanding of vegetation removal procedure is limited, meaning that the SRTM DEM is at present not available as a bare ground DTM. The future study focuses on the new vegetation remove method which requires global vegetation height data (Lefsky, 2010; Simard et al., 2011) on global DEM (e.g. SRTM). In particular, the percentage of canopy height that is removed from SRTM DEM should be quantified, as this percentage tends to differ from river to river. The improved performance of inundation modelling based on vegetation removed SRTM topography should be demonstrated as well.

*Data fusion towards a better topography for inundation modelling*

It is clear that SRTM topography needs to be improved to better support inundation modelling. Data fusion provides the new avenue to replace SRTM elevation with high vertical accuracy data, in such a way that improved inundation modelling could be performed on the hybrid/fused DEM. However, the high accuracy data are not often available, and usually costly. Ground and air survey (e.g. LiDAR) can be undertaken in few areas which are significant to inundation processes, improving modelling accuracy while limiting the costs. The question therefore arises from where and to what extent the new measurements should be undertaken, to improve inundation modelling in a satisfactory way, meanwhile, with lowest cost. In another word, where is the important location of the topography that dominates the flood inundation pattern, which badly needs to be improved? The future study should contain the following: Firstly, the topography comparison is carried out to identify the error areas on floodplain. Secondly, the selected areas (possibility with several combinations) are fused with data abstracted from high-resolution, high-accuracy LiDAR data. To assess the model performances, the inundation models based on those fused topographies are benchmarked with the LiDAR-based model.

*DEM sampling for random noise assessment*

SAR-derived DEMs (e.g. SRTM) suffer from random noise in the form of spikes and wells. Very few studies have removed the random noises from SRTM before integrating it into hydraulic modelling. The impact of the random noise in the topographic data to inundation modelling is unknown and yet to be explored. It would be interesting to study reconstructing the random noise by sampling the vertical elevation of cells in a high-resolution, high-accuracy LiDAR DEM. The sampling cell is randomly selected from the DEM domain. With various combinations of vertical elevation and location of the cells, the reconstructed DEMs are feed into inundation models to simulate a real-life flood event. Those models are benchmarked with the original high-resolution, high-accuracy LiDAR DEM.

*The integration of remote sensing data from the future missions with flood modelling*

The upcoming satellite missions will offer much more detailed and accurate information of topography, flood extent and water level globally, which most likely would provide huge value for flood modelling. As mentioned, these data include new global DEM (e.g. TanDEM-X), satellite imagery (e.g. Sentinel-1) and water levels (e.g. SWOT). The exploration of the potential and limitations of those data in flood modelling is of great interest of scientists and practitioners in this field. The methodological approaches developed in this thesis are general enough and could be very well adopted to this end.

# REFERENCES

Alsdorf DE, Dunne T, Melack J, Smith L, Hess L. 2005. Diffusion modelling of recessional flow on central Amazonian floodplains. *Geophysical Research Letters* **32**(21): L21405. doi: 10.1029/2005GL024412

Alsdorf DE, Rodríguez E, Lettenmaier DP. 2007. Measuring surface water from space. *Reviews of Geophysics* **45**(2). doi: 10.1029/2006rg000197

Alfieri L, Salamon P, Bianchi A, Neal J, Bates P, Feyen L. 2014. Advances in pan-European flood hazard mapping. *Hydrological Processes* **28**(13): 4067-4077. doi: 10.1002/hyp.9947

Aronica G, Bates PD, Horritt MS. 2002. Assessing the uncertainty in distributed model predictions using observed binary pattern information within GLUE. *Hydrological Processes* **16** (10): 2001–2016. doi: 10.1002/hyp.398

Aronica G, Hankin B, Beven K. 1998. Uncertainty and equifinality in calibrating distributed roughness coefficients in a flood propagation model with limited data. *Advances in Water Resources* **22**(4): 349-365.

Andreadis KM, Clark EA, Lettenmaier DP, Alsdorf DE. 2007. Prospects for river discharge and depth estimation through assimilation of swath-altimetry into a raster-based hydrodynamics model. *Geophysical Research Letters* **34**(10): L10403. doi: 10.1029/2007gl029721

Andreadis KM, Schumann GJP, Pavelsky T. 2013. A simple global river bankfull width and depth database. *Water Resources Research* **49**(10): 7164-7168. doi: 10.1002/wrcr.20440

Andreadis KM, Schumann GJP. 2014. Estimating the impact of satellite observations on the predictability of large-scale hydraulic models. *Advances in Water Resources* **73**, 44–54.

Airbus Defence & Space. 2014. Airbus Defence and Space, TerraSAR-X Services International Price List. February, 2014.

Bates PD, Neal JC, Alsdorf D, Schumann GJP. 2013. Observing Global Surface Water Flood Dynamics. *Surveys in Geophysics* **35**(3): 839-852. doi: 10.1007/s10712-013-9269-4

Bates PD. 2012. Integrating remote sensing data with flood inundation models: how far have we got? *Hydrological Processes* **26**(16): 2515-2521. doi: 10.1002/hyp.9374

Bates PD, Horritt MS, Aronica G, Beven K. 2004. Bayesian updating of flood inundation likelihoods conditioned on flood extent data. *Hydrological Processes* **18**(17): 3347-3370. doi: 10.1002/hyp.1499

Bates PD, Wilson MD, Horritt MS, Mason DC, Holden N, Currie A. 2006. Reach scale floodplain inundation dynamics observed using airborne synthetic aperture radar imagery: Data analysis and modelling. *Journal of Hydrology* **328**(1-2): 306-318. doi: 10.1016/j.jhydrol.2005.12.028

Bates PD 2004. Remote sensing and flood inundation modelling. *Hydrological Processes* **18**(13): 2593-2597. doi: 10.1002/hyp.5649

Bates PD, De Roo APJ 2000. A simple raster-based model for flood inundation simulation. *Journal of Hydrology* **236**(1-2): 54-77.

Bates PD, Horritt MS, Fewtrell TJ. 2010. A simple inertial formulation of the shallow water equations for efficient two-dimensional flood inundation modelling. *Journal of Hydrology* **387**(1-2): 33-45. doi: 10.1016/j.jhydrol.2010.03.027

Baugh CA, Bates PD, Schumann G, Trigg MA. 2013. SRTM vegetation removal and hydrodynamic modeling accuracy. *Water Resources Research* **49**(9): 5276-5289. doi: 10.1002/wrcr.20412

Balsamo G., et al., 2012. ERA-Interim/Land: A global land-surface reanalysis based on ERA-Interim meteorological forcing. ERA-Report series, n. 13, pp. 25

Beven KJ, Binley A. 1992. The future of distributed models: Model calibration and uncertainty prediction. *Hydrological Processes* **6**: 279-298. doi: 10.1002/hyp.3360060305

Beven, K. 2006 A manifesto for the equifinality thesis. *Journal of Hydrology.* **320**(1–2), 18–36.

Beven K, Smith P, Freer J. 2007 Comment on "Hydrological forecasting uncertainty assessment: incoherence of the GLUE methodology", by Mantovan P, Todini E. *Journal of Hydrology* 338, 315–318.

Beven K, Smith P, Freer J. 2008 So just why would a modeler choose to be incoherent? *Journal of Hydrology.* 354, 15–32.

Beven K, Freer J. 2001. Equifinality, data assimilation, and uncertainty estimation in mechanistic modelling of complex environmental systems using the GLUE methodology. *Journal of Hydrology* **249**(1–4): 11-29.

Brivio PA, Colombo R, Maggi M, Tomasoni R. 2002. Integration of remote sensing data and GIS for accurate mapping of flooded areas. *International Journal of Remote Sensing* **23**(3): 429-441. doi: 10.1080/01431160010014729

Bjerklie DM, Lawrence Dingman S, Vorosmarty CJ, Bolster CH, Congalton RG. 2003. Evaluating the potential for measuring river discharge from space. *Journal of Hydrology* **278**(1-4): 17-38. doi: 10.1016/s0022-1694(03)00129-x

Birkett CM. 1995. The contribution of TOPEX/POSEIDON to the global monitoring of climatically sensitive lakes. *Journal of Geophysical Research: Oceans* **100**(C12): 25179-25204. doi: 10.1029/95jc02125

Birkett CM 1998. Contribution of the TOPEX NASA Radar Altimeter to the global monitoring of large rivers and wetlands. *Water Resources Research* **34**(5): 1223-1239.

Birkett CM. 2002. Surface water dynamics in the Amazon Basin: Application of satellite radar altimetry. *Journal of Geophysical Research* **107**(D20). doi: 10.1029/2001jd000609

Birkinshaw SJ, O'Donnell GM, Moore P, Kilsby CG, Fowler HJ, Berry PAM. 2010. Using satellite altimetry data to augment flow estimation techniques on the Mekong River. *Hydrological Processes* **24**(26): 3811-3825. doi: 10.1002/hyp.7811

Biancamaria S, Bates PD, Boone A, Mognard NM. 2009. Large-scale coupled hydrologic and hydraulic modelling of the Ob river in Siberia. *Journal of Hydrology* **379**(1-2): 136-150. doi: 10.1016/j.jhydrol.2009.09.054

Brandimarte L, Brath A, Castellarin A, Baldassarre GD. 2009. Isla Hispaniola: A trans-boundary flood risk mitigation plan. *Physics and Chemistry of the Earth, Parts A/B/C* **34**(4–5): 209-218. doi: 10.1016/j.pce.2008.03.002

Brandimarte L, Di Baldassarre G. 2012. Uncertainty in design flood profiles derived by hydraulic modelling. *Hydrology Research* **43**(6): 753-761.

Brown CG, Sarabandi K, Pierce LE. 2010. Model-Based Estimation of Forest Canopy Height in Red and Austrian Pine Stands Using Shuttle Radar Topography Mission and Ancillary Data: A Proof-of-Concept Study. Geoscience and Remote Sensing, *IEEE Transactions on* **48**(3): 1105-1118. doi: 10.1109/tgrs.2009.2031635

Brath A, Castellarin A, Franchini M, Galeati G. 2001. Estimating the index flood using indirect methods. *Hydrological Sciences Journal* **46** (3), 399-418, doi:10.1080/02626660109492835.

Castellarin A. 2007. Probabilistic envelope curves for design flood estimation at ungauged sites. *Water Resources Research* **43** (4), W04406

Castellarin A, Di Baldassarre G, Bates P, Brath A. 2009. Optimal cross-sectional spacing in Preissmann Scheme 1D hydrodynamic models. *Journal of Hydraulic Engineering* **135**(2), 96–105.

Castellarin A, Di Baldassarre G, Brath A. 2011. Floodplain management strategies for flood attenuation in the River Po. *River Research Application* **27**(8), 1037–1047.

Charlton ME, Large ARG, Fuller IC. 2003. Application of airborne LiDAR in river environments: the River Coquet, Northumberland, UK. *Earth Surface Process and Landforms.* **28**(3), 299–306.

Calmant S, Lee H, Souza AE, Shum CK, Seyler F, Huang Z. 2009. JASON-2 IGDRs for flood alert in the Amazon Basin. *Ocean Surface Topography Science Team Meeting*, OSTST 2009, Seattle, WA, USA, 22–24 June 2009.

Coe MT, Costa MH, Howard EA. 2008. Simulating the surface waters of the Amazon River basin: impacts of new river geomorphic and flow parameterizations. *Hydrological Processes* **22**(14): 2542-2553. doi: 10.1002/hyp.6850

Covello F, Battazza F, Coletta A, Lopinto E, Fiorentino C, Pietranera L, Zoffoli S. 2010. COSMO-SkyMed an existing opportunity for observing the Earth. *Journal of Geodynamics* **49**(3–4): 171-180.

Chow, V.T. 1959 *Open Channel Hydraulics.* McGraw-Hill Book Co, New York.

Coratza, L. 2005 *Aggiornamento del Catasto delle Arginature Maestre di Po.* River Po Basin Authority, Parma.

Di Baldassarre G, Uhlenbrook S. 2012. Is the current flood of data enough? A treatise on research needs for the improvement of flood modelling. *Hydrological Processes* **26**(1): 153-158. doi: 10.1002/hyp.8226

Di Baldassarre G, Schumann G, Bates PD, Freer JE, Beven KJ. 2010. Flood-plain mapping: a critical discussion of deterministic and probabilistic approaches. *Hydrological Sciences Journal* **55**(3): 364-376. doi: 10.1080/02626661003683389

Di Baldassarre G, Schumann G, Brandimarte L, Bates P. 2011. Timely Low Resolution SAR Imagery To Support Floodplain Modelling: a Case Study Review. *Surveys in Geophysics* **32**(3): 255-269. doi: 10.1007/s10712-011-9111-9

Di Baldassarre G, Montanari A. 2009. Uncertainty in river discharge observations: a quantitative analysis. *Hydrology and Earth System Sciences* **13**(6): 913-921. doi: 10.5194/hess-13-913-2009

Di Baldassarre G, Schumann G, Bates P. 2009a. Near real time satellite imagery to support and verify timely flood modelling. *Hydrological Processes* **23**(5): 799-803. doi: 10.1002/hyp.7229

Di Baldassarre G, Schumann G, Bates PD. 2009b. A technique for the calibration of hydraulic models using uncertain satellite observations of flood extent. *Journal of Hydrology* **367**(3-4): 276-282. doi: 10.1016/j.jhydrol.2009.01.020

Di Baldassarre G, Castellarin A, Montanari A, Brath A. 2009c. Probability weighted hazard maps for comparing different flood risk management strategies: a case study. *Natural Hazards* doi: 10.1007/s11069-009-9355-6

Di Baldassarre, G., Laio, F. & Montanari, A. 2009d Design flood estimation using model selection criteria. *Phys. Chem. Earth, Parts A/B/C* **34**(10–12), 606–611.

Di Giuseppe F, Molteni F, Dutra E. 2013. Real-time correction of ERA-Interim monthly rainfall. *Geophysical Research Letters* **40**(14): 3750-3755. doi: 10.1002/grl.50670

Domeneghetti A, Tarpanelli A, Brocca L, Barbetta S, Moramarco T, Castellarin A, Brath A. 2014. The use of remote sensing-derived water surface data for hydraulic model calibration. *Remote Sensing of Environment* **149**: 130-141. doi: 10.1016/j.rse.2014.04.007

Dottori F, Di Baldassarre G, Todini E. 2013. Detailed data is welcome, but with a pinch of salt: Accuracy, precision, and uncertainty in flood inundation modeling. *Water Resources Research* **49**(9): 6079-6085. doi: 10.1002/wrcr.20406

de Paiva RCD, Buarque DC, Collischonn W, Bonnet M-P, Frappart F, Calmant S, Bulhões Mendes CA. 2013. Large-scale hydrologic and hydrodynamic modeling of the Amazon River basin. *Water Resources Research* **49**(3): 1226-1243. doi: 10.1002/wrcr.20067

de Moel H, van Alphen J, Aerts JCH. 2009. Flood maps in Europe - methods, availability and use. *Natural Hazards and Earth System Sciences* **9**(2): 289-301.

Eineder M, Fritz T, Jaber W, Rossi C, Breit H, Decadal Earth Topography Dynamics Measured with TanDEM-X and SRTM. *Proceedings of IGARSS (International Geoscience and Remote Sensing Symposium)*, Munich, Germany, July 22-27, 2012

e-GEOS. 2014. Price list. January 2014

ElMonshid BEF, ElAwad OMA, Ahmed SE. 1997. Environmental effect of the Blue Nile Sediment on reservoirs and Irrigation Canals. International 5th Nile 2002 Conference, Addis Ababa, Ethiopia.

Fabio P, Aronica GT, Apel H. 2010. Towards automatic calibration of 2-D flood propagation models. *Hydrology and Earth System Sciences* **14**(6): 911–924.

Falorni G, Teles V, Vivoni ER, Bras RL, Amaratunga KS. 2005. Analysis and characterization of the vertical accuracy of digital elevation models from the Shuttle Radar Topography Mission. *Journal of Geophysical Research: Earth Surface* **110**(F2): F02005. doi: 10.1029/2003jf000113

Farr TG, Rosen PA, Caro E, Crippen R, Duren R, Hensley S, . . . Alsdorf D. 2007. The Shuttle Radar Topography Mission. *Reviews of Geophysics* **45**(2). doi: 10.1029/2005rg000183

Fewtrell TJ, Bates PD, Horritt M, Hunter NM. 2008. Evaluating the effect of scale in flood inundation modelling in urban environments. *Hydrological Processes* **22**(26): 5107-5118. doi: 10.1002/hyp.7148

Frappart F, Calmant S, Cauhopé M, Seyler F, Cazenave A. 2006. Preliminary results of ENVISAT RA-2-derived water levels validation over the Amazon basin. *Remote Sensing of Environment* **100**(2): 252-264.

Giustarini L, Matgen P, Hostache R, Montanari M, Plaza D, Pauwels VRN, . . . Savenije HHG. 2011. Assimilating SAR-derived water level data into a hydraulic model: a case study. *Hydrol. Earth Syst. Sci.* **15**(7): 2349-2365. doi: 10.5194/hess-15-2349-2011

Giustarini L, Hostache R, Matgen P, Schumann GJP, Bates PD, Mason DC. 2013. A Change Detection Approach to Flood Mapping in Urban Areas Using TerraSAR-X. *Geoscience and Remote Sensing, IEEE Transactions on* **51**(4): 2417-2430. doi: 10.1109/tgrs.2012.2210901

Gichamo TZ, Popescu I, Jonoski A, Solomatine D. 2012. River cross-section extraction from the ASTER global DEM for flood modeling. *Environmental Modelling & Software* **31**: 37-46. doi: 10.1016/j.envsoft.2011.12.003

Getirana ACV, Bonnet M-P, Calmant S, Roux E, Rotunno Filho OC, Mansur WJ. 2009. Hydrological monitoring of poorly gauged basins based on rainfall–runoff modeling and spatial altimetry. *Journal of Hydrology* **379**(3–4): 205-219. doi: http://dx.doi.org/10.1016/j.jhydrol.2009.09.049

Gruber A, Wessel B, Huber M, Breunig M, Wagenbrenner S, Roth A. 2012. Quality assessment of first TanDEM-XDEMs for different terrain types. *9th European Conference on Synthetic Aperture Radar Program*

García-Pintado J, Neal JC, Mason DC, Dance SL, Bates PD. 2013 Scheduling satellite-based SAR acquisition for sequential assimilation of water level observations into flood modelling. *Journal of Hydrology* 495, 252–266.

Hostache R, Lai X, Monnier J, Puech C. 2010. Assimilation of spatially distributed water levels into a shallow-water flood model. Part II: Use of a remote sensing image of Mosel River. *Journal of Hydrology* **390**(3–4): 257-268. doi: http://dx.doi.org/10.1016/j.jhydrol.2010.07.003

Hanssen R. 2001. *Radar Interferometry: Data Interpretation and Error Analysis.* Kluwer Academic Publishers, Springer, Dordrecht, the Netherlands, 308 p.

Hannah DM, Demuth S, van Lanen HAJ, Looser U, Prudhomme C, Rees G, Tallaksen LM. 2011. Large-scale river flow archives: importance, current status and future needs. *Hydrological Processes* **25**(7): 1191-1200. doi: 10.1002/hyp.7794

Hall AC, Schumann GJP, Bamber JL, Bates PD. 2011. Tracking water level changes of the Amazon Basin with space-borne remote sensing and integration with large scale hydrodynamic modelling: A review. *Physics and Chemistry of the Earth, Parts A/B/C* **36**(7-8): 223-231. doi: 10.1016/j.pce.2010.12.010

Hess LL, Melack JM, Filoso S, Yong W. 1995. Delineation of inundated area and vegetation along the Amazon floodplain with the SIR-C synthetic aperture radar. *Geoscience and Remote Sensing, IEEE Transactions on* **33**(4): 896-904. doi: 10.1109/36.406675

Horritt MS. 1999. A statistical active contour model for SAR image segmentation. *Image and Vision Computing* **17**(3-4): 213-224.

Horritt MS, Bates PD. 2001. Effects of spatial resolution on a raster based model of flood flow. *Journal of Hydrology* **253**(1-4): 239-249.

Horritt MS, Bates PD. 2001. Predicting floodplain inundation: raster-based modelling versus the finite-element approach. *Hydrological Processes* **15**(5): 825-842. doi: 10.1002/hyp.188

Horritt MS. 2000. Calibration of a two-dimensional finite element flood flow model using satellite radar imagery. *Water Resources Research* **36**(11): 3279-3291. doi: 10.1029/2000wr900206

Horritt MS. 2006. A methodology for the validation of uncertain flood inundation models. *Journal of Hydrology* **326**(1-4): 153-165.

Horritt MS, Di Baldassarre G, Bates PD, Brath A. 2007. Comparing the performance of a 2-D finite element and a 2-D finite volume model of floodplain inundation using airborne SAR imagery. *Hydrological Processes* **21**(20): 2745-2759. doi: 10.1002/hyp.6486

Horritt MS, Bates PD. 2002. Evaluation of 1D and 2D numerical models for predicting river flood inundation. *Journal of Hydrology* **268**(1–4): 87-99. doi: http://dx.doi.org/10.1016/S0022-1694(02)00121-X

Horritt MS, Bates PD, Fewtrell TJ, Mason DC, Wilson MD. 2010. Modelling the hydraulics of the Carlisle 2005 flood event. *Proceedings of the ICE Water Management* **163**(6), 273–281.

Hostache R, Matgen P, Schumann G, Puech C, Hoffmann L, Pfister L. 2009. Water Level Estimation and Reduction of Hydraulic Model Calibration Uncertainties Using Satellite SAR Images of Floods. *Geoscience and Remote Sensing, IEEE Transactions on* **47**(2): 431-441. doi: 10.1109/tgrs.2008.2008718

Hunter NM, Bates PD, Horritt MS, Wilson MD. 2007. Simple spatially-distributed models for predicting flood inundation: A review. *Geomorphology* **90**(3-4): 208-225.

Hunter NM, Horritt MS, Bates PD, Wilson MD, Werner MG. 2005. An adaptive time step solution for raster-based storage cell modelling of floodplain inundation. *Advances in Water Resources* **28**(9): 975-991.

Hunter NM, Bates PD, Horritt MS, De Roo APJ, Werner MGF. 2005. Utility of different data types for calibrating flood inundation models within a GLUE framework. *Hydrology and Earth System Sciences* **9**(4). doi: 10.5194/hess-9-412-2005

Huffman GJ, Adler RF, Bolvin DT, Gu G. 2009. Improving the global precipitation record: GPCPVersion 2.1. *Geophysical Research Letters* **36**: L17808, doi:10.1029/2009gl040000.

Hrachowitz M, Savenije HHG, Blöschl G, McDonnell JJ, Sivapalan M, Pomeroy JW, et al. 2013. A decade of Predictions in Ungauged Basins (PUB)—a review. *Hydrological Sciences Journal* **58**(6): 1198-1255.

Hwang C, Peng M-F, Ning J, Luo J, Sui C-H. 2005. Lake level variations in China from TOPEX/Poseidon altimetry: data quality assessment and links to precipitation and ENSO. *Geophysical Journal International* **161**(1): 1-11. doi: 10.1111/j.1365-246X.2005.02518.x

Imhoff ML, Vermillion CH, Story M, Polcyn F. 1986. Space-borne radar for monsoon and storm induced flood control planning in Bangladesh: A result of the shuttle imaging radar-B program. *Science of The Total Environment* **56**(0): 277-286. doi: http://dx.doi.org/10.1016/0048-9697(86)90332-3

IAHS. 2003. World Catalogue of Maximum Observed Floods, IAHS Publication 284.

Jarvis A, Reuter HI, Nelson A, Guevara E. 2008. Hole-filled SRTM for the globe Version 4, available from the CGIAR-CSI SRTM 90m Database (http://srtm.csi.cgiar.org).

Jung Y, Merwade V. 2012a. Uncertainty Quantification in Flood Inundation Mapping Using Generalized Likelihood Uncertainty Estimate and Sensitivity Analysis. *Journal of Hydrologic Engineering* **17**(4): 507-520. doi: 10.1061/(asce)he.1943-5584.0000476

Jung HC, Jasinski M, Kim J-W, Shum CK, Bates P, Neal J, . . . Alsdorf D. 2012b. Calibration of two-dimensional floodplain modeling in the central Atchafalaya Basin Floodway System using SAR interferometry. *Water Resources Research* **48**(7): W07511. doi: 10.1029/2012wr011951

Kayastha N, Shrestha DL, Solomatine, DP. 2010. Experiments with several methods of parameter uncertainty estimation in hydrological modeling. In: *Proc. 9th International Conference on Hydroinformatics*, September, Tianjin, China.

Koutsoyiannis D. 2014. Hydrology, society, change and uncertainty. *Geophysical Research Abstracts* **16**: EGU2014-4243.

Koblinsky CJ, Clarke RT, Brenner AC, Frey H. 1993. Measurement of river level variations with satellite altimetry. *Water Resources Research* **29**(6): 1839-1848. doi: 10.1029/93wr00542

Kouraev AV, Zakharova EA, Samain O, Mognard NM, Cazenave A. 2004. Ob' river discharge from TOPEX/Poseidon satellite altimetry (1992–2002). *Remote Sensing of Environment* **93**(1–2): 238-245. doi: http://dx.doi.org/10.1016/j.rse.2004.07.007

Krieger G, Zink M, Fiedler H, Hajnsek I, Younis M, Huber S, ..... THE TanDEM-X Mission: Overview and status, Radar Conference, 2009 IEEE , vol., no., pp.1,5, 4-8 May 2009. doi: 10.1109/RADAR.2009.4977075

Krueger T, Freer J, Quinton JN, Macleod CJA, Bilotta GS, Brazier RE, . . . Haygarth PM. 2010. Ensemble evaluation of hydrological model hypotheses. *Water Resources Research* **46**(7): W07516. doi: 10.1029/2009wr007845

LeFavour G. 2005. Water slope and discharge in the Amazon River estimated using the shuttle radar topography mission digital elevation model. *Geophysical Research Letters* **32**(17). doi: 10.1029/2005gl023836

Lehner B, Verdin K, Jarvis A. 2008. New Global Hydrography Derived From Spaceborne Elevation Data. Eos, *Transactions American Geophysical Union* **89**(10): 93-94. doi: 10.1029/2008eo100001

Lefsky MA. 2010. A global forest canopy height map from the Moderate Resolution Imaging Spectroradiometer and the Geoscience Laser Altimeter System. *Geophysical Research Letters* **37**(15): L15401. doi: 10.1029/2010gl043622

Leon JG, Calmant S, Seyler F, Bonnet MP, Cauhopé M, Frappart F, . . . Fraizy P. 2006. Rating curves and estimation of average water depth at the upper Negro River based on satellite altimeter data and modeled discharges. *Journal of Hydrology* **328**(3-4): 481-496. doi: 10.1016/j.jhydrol.2005.12.006

Leopold LB, Maddock, TJ. 1953. The hydraulic geometry of stream channels and some physiographic implications. *U.S. Geological Survey Professional Paper* **252**, 56

Liu Y, Nishiyama S, Yano T. 2004. Analysis of four change detection algorithms in bi-temporal space with a case study. *International Journal of Remote Sensing* **25**(11): 2121-2139. doi: 10.1080/01431160310001606647

Lu D, Mausel P, Brondízio E, Moran E. 2004. Change detection techniques. *International Journal of Remote Sensing* **25**(12): 2365-2401. doi: 10.1080/0143116031000139863

Ludwig R, Schneider P. 2006. Validation of digital elevation models from SRTM X-SAR for applications in hydrologic modeling. *ISPRS Journal of Photogrammetry and Remote Sensing* **60**(5), 339–358.

Mantovan P, Todini E. 2006. Hydrological forecasting uncertainty assessment: Incoherence of the GLUE methodology. *Journal of Hydrology* **330**(1–2): 368-381.

Maione U, Mignosa P, Tomirotti M. 2003. Regional estimation model of synthetic design hydrographs. *International Journal of River Basin Management* **12**, 151–163.

Maheu C, Cazenave A, Mechoso CR. 2003. Water level fluctuations in the Plata Basin (South America) from Topex/Poseidon Satellite Altimetry. *Geophysical Research Letters* **30**(3): 1143. doi: 10.1029/2002gl016033

Marcus WA, Fonstad MA. 2008. Optical remote mapping of rivers at sub-meter resolutions and watershed extents. *Earth Surface Processes and Landforms* **33**(1): 4-24. doi: 10.1002/esp.1637

Mason DC, Bates PD, Dall' Amico JT. 2009. Calibration of uncertain flood inundation models using remotely sensed water levels. *Journal of Hydrology* **368**(1-4): 224-236. doi: 10.1016/j.jhydrol.2009.02.034

Mason DC, Schumann GJP, Neal JC, Garcia-Pintado J, Bates PD. 2012. Automatic near real-time selection of flood water levels from high resolution Synthetic Aperture Radar images for assimilation into hydraulic models: A case study. *Remote Sensing of Environment* **124**(0): 705-716. doi: http://dx.doi.org/10.1016/j.rse.2012.06.017

Mason DC, Giustarini L, Garcia-Pintado J, Cloke HL. 2014. Detection of flooded urban areas in high resolution Synthetic Aperture Radar images using double scattering. International *Journal of Applied Earth Observation and Geoinformation* **28**(0): 150-159. doi: http://dx.doi.org/10.1016/j.jag.2013.12.002

Matgen P, Hostache R, Schumann G, Pfister L, Hoffmann L, Savenije HHG. 2011. Towards an automated SAR-based flood monitoring system: Lessons learned from two case studies. *Physics and Chemistry of the Earth, Parts A/B/C* **36**(7–8): 241-252. doi: http://dx.doi.org/10.1016/j.pce.2010.12.009

Matgen P, Henry JB, Pappenberger F, De Fraipont P, Hoffmann L, Pfister L. 2004. Uncertainty in calibrating flood propagation models with flood boundaries observed from synthetic aperture radar imagery. 20th ISPRS Congress, International Society for Photogrammetry and Remote Sensing, Istanbul.

Matgen P, Schumann G, Henry JB, Hoffmann L, Pfister L. 2007a. Integration of SAR-derived river inundation areas, high-precision topographic data and a river flow model toward near real-time flood management. *International Journal of Applied Earth Observation and Geoinformation* **9**(3): 247-263. doi: 10.1016/j.jag.2006.03.003

Matgen P, Schumann G, Pappenberger F, Pfister L. 2007b. Sequential assimilation of remotely sensed water stages in flood inundation models, in Symposium HS3007 at IUGG2007: Remote Sensing for Environmental Monitoring and Change Detection, IAHS Publication, 316, 78–88.

Matgen P, Montanari M, Hostache R, Pfister L, Hoffmann L, Plaza D, . . . Savenije HHG. 2010. Towards the sequential assimilation of SAR-derived water stages into hydraulic models using the Particle Filter: proof of concept. *Hydrological and Earth System Sciences* **14**(9): 1773-1785. doi: 10.5194/hess-14-1773-2010

Mazzoleni M, Bacchi B, Barontini S, Di Baldassarre G, Pilotti M, Ranzi R. 2014. Flooding Hazard Mapping in Floodplain Areas Affected by Piping Breaches in the Po River, Italy. *Journal of Hydrologic Engineering* **19**(4): 717-731. doi: doi:10.1061/(ASCE)HE.1943-5584.0000840

Martinis, S, Twele, A, Voigt, S. 2009. Towards operational near-real time flood detection using a split-based automatic thresholding procedure on high resolution TerraSAR-X data. *Natural Hazards and Earth System Science* **9**, 303–314.

Martinis S, Twele A, Strobl C, Kersten J, Stein E. 2013. A Multi-Scale Flood Monitoring System Based on Fully Automatic MODIS and TerraSAR-X Processing Chains. *Remote Sensing* **5**(11): 5598-5619. doi:10.3390/rs5115598

Martinis S, Kersten J, Twele A. A fully automated TerraSAR-X based flood service. *ISPRS Journal of Photogrammetry and Remote Sensing*

MacIntosh H, Profeti G. 1995. The use of ERS SAR data to manage flood emergencies at the smaller scale. 2nd ERS Applications Workshop, European Space Agency, London.

Merz B, Thieken AH, Gocht M. 2007. Flood risk mapping at the local scale: concepts and challenges. In: *Flood Risk Management in Europe: Innovation in Policy and Practice,* Begum S, Stive MJF, Hall JW. (eds). Advances in Natural and Technological Hazards Research Series no. 25, Springer, Dordrecht, The Netherlands. Ch. 13, 231–251.

Merz R, Blöschl G. 2008a. Flood frequency hydrology: 1. Temporal, spatial, and causal expansion of information. *Water Resources Research* **44**(8), W08432, doi:10.1029/2007wr006744.

Merz R, Blöschl G. 2008. Flood frequency hydrology: 2. Combining data evidence. *Water Resources Research* **44**(8), W08433, doi:10.1029/2007wr006745.

Montanari A. 2005. Large sample behaviors of the generalized likelihood uncertainty estimation (GLUE) in assessing the uncertainty of rainfall-runoff simulations. *Water Resources Research* **41**(8): W08406.

Montanari A. 2007. What do we mean by 'uncertainty'? The need for a consistent wording about uncertainty assessment in hydrology. *Hydrological Processes* **21**(6): 841-845. doi: 10.1002/hyp.6623

Moore P., et al. 2014. CRUCIAL: Cryosat-2 Success over Inland Water and Land. Geophysical Research Abstracts (ISSN: 1607-7962), 16, 2014

Merwade V. 2008. Uncertainty in flood inundation mapping: current issues and future directions. *Journal of Hydrological Engineering* **13**(7), 608–620.

Moya QV, Popescu I, Solomatine DP, Bociort L. 2012. Cloud and cluster computing in uncertainty analysis of integrated flood models. *Journal of Hydroinformatics* doi:10.2166/hydro.2012.017

Milzow C, Krogh PE, Bauer-Gottwein P. 2011. Combining satellite radar altimetry, SAR surface soil moisture and GRACE total storage changes for hydrological model calibration in a large poorly gauged catchment. *Hydrology and Earth System Sciences* **15**(6): 1729-1743.

Mukolwe M, Yan K, Di Baldassarre G, Solomatine D. 2015. Testing new sources of topographic data for flood propagation modelling under structural, parameter and observation uncertainty. *Hydrological Sciences Journal* doi: 10.1080/02626667.2015.1019507

Neal J, Schumann G, Bates P. 2012a. A subgrid channel model for simulating river hydraulics and floodplain inundation over large and data sparse areas. *Water Resources Research* **48**(11): W11506. doi: 10.1029/2012wr012514

Neal J, Villanueva I, Wright N, Willis T, Fewtrell T, Bates P. 2012b. How much physical complexity is needed to model flood inundation? *Hydrological Processes* **26**(15): 2264-2282. doi: 10.1002/hyp.8339

Neal J, Bates PD, Fewtrell TJ, Hunter NM, Wilson MD, Horritt MS. 2009. Distributed whole city water level measurements from the Carlisle 2005 urban flood event and comparison with hydraulic model simulations. *Journal of Hydrology* **368**(1–4): 42-55. doi: http://dx.doi.org/10.1016/j.jhydrol.2009.01.026

Neal J, Keef C, Bates PD, Beven K, Leedal D. 2013. Probabilistic flood risk mapping including spatial dependence. *Hydrological Processes* **27**(9): 1349-1363.

Neal J, Schumann G, Bates P, Buytaert W, Matgen P, Pappenberger F. 2009. A data assimilation approach to discharge estimation from space. *Hydrological Processes* **23**(25): 3641-3649. doi: 10.1002/hyp.7518

Oberstadler R, HÖNsch H, Huth D. 1997. Assessment of the mapping capabilities of ERS-1 SAR data for flood mapping: a case study in Germany. *Hydrological Processes* **11**(10): 1415-1425.

Pappenberger F, Frodsham K, Beven K, Romanowicz R, Matgen P. 2007. Fuzzy set approach to calibrating distributed flood inundation models using remote sensing observations. *Hydrology and Earth System Sciences* **11**(2). doi: 10.5194/hess-11-739-2007

Pappenberger F, Matgen P, Beven KJ, Henry J-B, Pfister L, Fraipont de P. 2006. Influence of uncertain boundary conditions and model structure on flood inundation predictions. *Advances in Water Resources* **29**(10): 1430-1449. doi: http://dx.doi.org/10.1016/j.advwatres.2005.11.012

Pappenberger F, Beven K, Horritt M, Blazkova S. 2005. Uncertainty in the calibration of effective roughness parameters in HEC-RAS using inundation and downstream level observations. *Journal of Hydrology* **302**(1-4): 46-69. doi: 10.1016/j.jhydrol.2004.06.036

Pappenberger F, Matgen P, Beven KJ, Henry J-B, Pfister L, Fraipont de P. 2006. Influence of uncertain boundary conditions and model structure on flood inundation predictions. *Advances in Water Resources* **29**(10): 1430-1449. doi: http://dx.doi.org/10.1016/j.advwatres.2005.11.012

Pappenberger F, Beven KJ, Ratto M. Matgen P. 2008. Multi-method global sensitivity analysis of flood inundation models. *Advances in Water Resources* **31**, 1–14.

Pappenberger F, Dutra E, Wetterhall F, Cloke HL. 2012. Deriving global flood hazard maps of fluvial floods through a physical model cascade. *Hydrology and Earth System Sciences* **16**(11): 4143-4156. doi: 10.5194/hess-16-4143-2012

Padi PT, Baldassarre GD, Castellarin A. 2011. Floodplain management in Africa: Large scale analysis of flood data. *Physics and Chemistry of the Earth, Parts A/B/C* **36**(7-8): 292-298. doi: 10.1016/j.pce.2011.02.002

Patro S, Chatterjee C, Singh R, Raghuwanshi NS. 2009. Hydrodynamic modelling of a large flood-prone river system in India with limited data. *Hydrological Processes* **23**(19): 2774-2791. doi: 10.1002/hyp.7375

Papathanassiou KP, Cloude SR. 2001. Single-baseline polarimetric SAR interferometry. *Geoscience and Remote Sensing, IEEE Transactions on* **39**(11): 2352-2363. doi: 10.1109/36.964971

Paiva RCD, Collischonn W, Tucci CEM. 2011. Large scale hydrologic and hydrodynamic modeling using limited data and a GIS based approach. *Journal of Hydrology* **406**(3–4): 170-181.

Pereira-Cardenal SJ, Riegels ND, Berry PA, Smith M, Yakovlev RG, Siegfried A, Bauer-Gottwein P. 2011. Real-time remote sensing driven river basin modeling using radar altimetry. *Hydrology and Earth System Sciences* **15**(1): 241-254.

Petersen G, Fohrer N. 2010. Two-dimensional numerical assessment of the hydrodynamics of the Nile swamps in southern Sudan. *Hydrological Sciences Journal* **55**(1): 17-26. doi: 10.1080/02626660903525252

Pierdicca N, Pulvirenti L, Chini M, Guerriero L, Candela L. 2013. Observing floods from space: Experience gained from COSMO-SkyMed observations. *Acta Astronautica* **84**(0): 122-133.

Prestininzi P, Di Baldassarre G, Schumann G, Bates PD. 2011. Selecting the appropriate hydraulic model structure using low-resolution satellite imagery. *Advances in Water Resources* **34**(1): 38-46. doi: 10.1016/j.advwatres.2010.09.016

Preissmann A. 1961 Propagation of translatory waves in channels and rivers. In: *Proc. First Congress of French Association for Computation*, Grenoble, France.

Pulvirenti L, Pierdicca N, Chini M, Guerriero L. 2011. An algorithm for operational flood mapping from Synthetic Aperture Radar (SAR) data using fuzzy logic. *Nat. Hazards Earth Syst. Sci.* **11**(2): 529-540. doi: 10.5194/nhess-11-529-2011

Pulvirenti L, Chini M, Marzano FS, Pierdicca N, Mori S, Guerriero L, Boni G, Candela L. 2012. Detection of floods and heavy rain using Cosmo–SkyMed data: the event in Northwestern Italy of November 2011. IEEE International Geoscience and Remote Sensing Symposium (IGARSS 2012), Munich, 22–27 July, pp. 3026–3029.

Rodriguez E, Morris CS, Belz JE. 2006. A Global Assessment of the SRTM Performance. *Photogrammetric Engineering and Remote Sensing* **72**(3): 249-260.

Rabus B, Eineder M, Roth A, Bamler R. 2003. The shuttle radar topography mission—a new class of digital elevation models acquired by spaceborne radar. *ISPRS Journal of Photogrammetry and Remote Sensing* **57**(4): 241-262. doi: 10.1016/s0924-2716(02)00124-7

Refsgaard JC, van der Sluijs JP, Højberg AL, Vanrolleghem PA. 2007. Uncertainty in the environmental modelling process – A framework and guidance. Environmental Modelling & Software 22(11): 1543-1556.

Romanowicz R, Beven K. 1998. Dynamic real-time prediction of flood inundation probabilities. *Hydrological Sciences Journal* **43**(2): 181-196.

Romanowicz R, Beven K. 2003. Estimation of flood inundation probabilities as conditioned on event inundation maps. *Water Resources Research* **39**(3), 1073.

Samuels P. 1989. Backwater length in rivers. *ICE Proceedings* **87**: 571-582.

Sanders BF. 2007. Evaluation of on-line DEMs for flood inundation modeling. *Advances in Water Resources* **30**(8): 1831-1843.

Santos da Silva J, Calmant S, Seyler F, Rotunno Filho OC, Cochonneau G, Mansur WJ. 2010. Water levels in the Amazon basin derived from the ERS 2 and ENVISAT radar altimetry missions. *Remote Sensing of Environment* **114**(10): 2160-2181. doi: 10.1016/j.rse.2010.04.020

Sanyal J, Lu XX. 2004. Application of Remote Sensing in Flood Management with Special Reference to Monsoon Asia: A Review. *Natural Hazards* **33**(2): 283-301. doi: 10.1023/b:nhaz.0000037035.65105.95

SARCOM. 2010. ERS/Envisat Products and Services, SARCOM International Price List. April, 2010.

Simard M, Pinto N, Fisher JB, Baccini A. 2011. Mapping forest canopy height globally with spaceborne lidar. *Journal of Geophysical Research: Biogeosciences* **116**(G4): G04021. doi: 10.1029/2011jg001708

Schumann G, Henry JB, Hoffmann L, Pfister L, Pappenberger F, Matgen P. 2005. Demonstrating the high potential of remote sensing in hydraulic modelling and flood risk management. *Annual Conference of the Remote Sensing and Photogrammetry Society With the NERC Earth Observation Conference*, Remote Sensing and Photogrammetry Society, Portsmouth, U. K.

Schumann G, Matgen P, Hoffmann L, Hostache R, Pappenberger F, Pfister L. 2007. Deriving distributed roughness values from satellite radar data for flood inundation modelling. *Journal of Hydrology* **344**(1-2): 96-111.

Schumann G, Di Baldassarre G, Bates PD. 2009. The Utility of Spaceborne Radar to Render Flood Inundation Maps Based on Multialgorithm Ensembles. *Geoscience and Remote Sensing, IEEE Transactions on* **47**(8): 2801-2807.

Schumann G, Neal JC, Phanthuwongpakdee K, Voisin N, Aspin T. 2012. Assessing forecast skill of a large scale 2D inundation model of the Lower Zambezi River with multiple satellite data sets. *XIX International Conference on Water Resources*. University of Illinois at Urbana-Champaign.

Schumann G, Di Baldassarre G, Alsdorf D, Bates PD. 2010. Near real-time flood wave approximation on large rivers from space: Application to the River Po, Italy. *Water Resources Research* **46**(5): W05601. doi: 10.1029/2008wr007672

Schumann G, Matgen P, Cutler MEJ, Black A, Hoffmann L, Pfister L. 2008. Comparison of remotely sensed water stages from LiDAR, topographic contours and SRTM. *ISPRS Journal of Photogrammetry and Remote Sensing* **63**(3): 283-296. doi: 10.1016/j.isprsjprs.2007.09.004

Schumann G, Bates PD, Horritt MS, Matgen P, Pappenberger F. 2009. Progress in integration of remote sensing–derived flood extent and stage data and hydraulic models. *Reviews of Geophysics* **47**(4). doi: 10.1029/2008rg000274

Schumann G, Di Baldassarre G. 2010. The direct use of radar satellites for event-specific flood risk mapping. *Remote Sensing Letters* **1**(2): 75-84. doi: 10.1080/01431160903486685

Schumann G, Bates PD, Neal JC, Andreadis KM. 2014 Measuring and mapping flood processes in *Hydro-meteorological Risks, Hazards, and Disasters*, Elsevier. In press

Schumann GJP, Neal JC, Voisin N, Andreadis KM, Pappenberger F, Phanthuwongpakdee N, Bates PD. 2013. A first large-scale flood inundation forecasting model. *Water Resources Research* **49**(10): 6248-6257. doi: 10.1002/wrcr.20521

Stokstad E. 1999. Scarcity of rain, stream gages threatens forecasts. *Science* **285**: 1199-1200.

Sivapalan M, Takeuchi K, Franks SW, Gupta VK, Karambiri H, Lakshmi V, . . . Zehe E. 2003. IAHS Decade on Predictions in Ungauged Basins (PUB), 2003–2012: Shaping an exciting future for the hydrological sciences. *Hydrological Sciences Journal* **48**(6): 857-880. doi: 10.1623/hysj.48.6.857.51421

Smith LC. 1997. Satellite remote sensing of river inundation area, stage, and discharge: a review. *Hydrological Processes* **11**(10): 1427-1439.

Stephens E, Schumann G, Bates P. 2013. Problems with binary pattern measures for flood model evaluation. *Hydrological Processes* **28**(18): 4928-4937.

Sutcliffe J.V., and Parks, Y.P., 1999. The hydrology of the Nile. IAHS Special Publication no.

Sweet WV, Geratz JW. 2003. Bankfull hydraulic geometry relationships and recurrence intervals for north carolina's coastal plain. *JAWRA Journal of the American Water Resources Association* **39**(4): 861-871. doi: 10.1111/j.1752-1688.2003.tb04411.x

Soergel U, Jacobsen K, Schack L. 2013 The TanDEM-X Mission: Data Collection and Deliverables. Hannover

Solomatine DP, Shrestha DL. 2009. A novel method to estimate model uncertainty using machine learning techniques. *Water Resources Research* **45**(12). doi: 10.1029/2008wr006839

Stedinger JR, Vogel RM, Lee SU, Batchelder R. 2008. Appraisal of the generalized likelihood uncertainty estimation (GLUE) method. *Water Resources Research* **44**(12): W00B06. doi: 10.1029/2008wr006822

Sun G, Ranson KJ, Kharuk VI, Kovacs K. 2003. Validation of surface height from shuttle radar topography mission using shuttle laser altimeter. *Remote Sensing Environment* **88**(4): 401–411.

Sun WC, Ishidaira H, Bastola S. 2010. Towards improving river discharge estimation in ungauged basins: calibration of rainfall-runoff models based on satellite observations of river flow width at basin outlet. *Hydrology and Earth System Sciences* **14**: 2011-2022.

Tachikawa T, Kaku M, Iwasaki A, Gesch D, Oimoen M, Zhang Z, Carabajal C. 2011. ASTER global digital elevation model version 2—summary of validation results. NASA, 27 pp.

Tarpanelli A, Brocca L, Melone F, Moramarco T. 2013. Hydraulic modelling calibration in small rivers by using coarse resolution synthetic aperture radar imagery. *Hydrological Processes* **27**(9): 1321-1330. doi: 10.1002/hyp.9550

Torres R, Snoeij P, Geudtner D, Bibby D, Davidson M, Attema E, . . . Rostan F. 2012. GMES Sentinel-1 mission. *Remote Sensing of Environment* 120(0): 9-24. doi: http://dx.doi.org/10.1016/j.rse.2011.05.028

US Army Corps of Engineers. 2001 Hydraulic Reference Manual. US Army Corps of Engineers, Hydrologic Engineering Center, Davis, CA, USA.

UNESCO. 1984. World Catalogue of Maximum Observed Floods, IAHS Publication 143.

Van Alphen J, Martini F, Loat R, Slomp R, Passchier, R. 2009. Flood risk mapping in Europe, experiences and best practices. *Journal of Flood Risk Management* **2**(4): 285-292.

Viglione A, Laio F, Claps P. 2007. A comparison of homogeneity tests for regional frequency analysis. *Water Resources Research* **43**(3), W03428.

Wagener T, Lees MJ, Wheater HS. 2001. A toolkit for the development and application of parsimonious hydrological models. *Mathematical Models of Large Watershed Hydrology* **1**: 87-136.

Wang Y, Liao M, Sun G, Gong J. 2005. Analysis of the water volume, length, total area and inundated area of the Three Gorges Reservoir, China using the SRTM DEM data. *International Journal of Remote Sensing* **26**(18): 4001-4012. doi: 10.1080/01431160500176788

Walker JP, Willgoose GR. 1999. On the effect of digital elevation model accuracy on hydrology and geomorphology. *Water Resources Research* **35**(7), 2259–2268.

Wachter K. 2007. The Analysis of the Danube Floods 2006. Vienna, Austria, ICPDR-International Commission for the Protection of the Danube River.

Westerhoff RS, Kleuskens MPH, Winsemius HC, Huizinga HJ, Brakenridge GR, Bishop C. 2013. Automated global water mapping based on wide-swath orbital synthetic-aperture radar. *Hydrological and Earth System Sciences* **17**, 651–663.

Wang W, Yang X, Yao T. 2012. Evaluation of ASTER GDEM and SRTM and their suitability in hydraulic modelling of a glacial lake outburst flood in southeast Tibet. *Hydrological Processes* **26**(2): 213-225. doi: 10.1002/hyp.8127

Wharton G, Arnell NW, Gregory KJ, Gurnell AM. 1989. River discharge estimated from channel dimensions. *Journal of Hydrology* **106**(3–4): 365-376. doi: http://dx.doi.org/10.1016/0022-1694(89)90080-2

Wilson M, Bates P, Alsdorf D, Forsberg B, Horritt M, Melack J, Famiglietti J. 2007. Modeling large-scale inundation of Amazonian seasonally flooded wetlands. *Geophysical Research Letters* **34**(15). doi: 10.1029/2007gl030156

Winsemius HC, Schaefli B, Montanari A, Savenije HHG. 2009. On the calibration of hydrological models in ungauged basins: A framework for integrating hard and soft hydrological information. Water Resources Research 45(12): W12422. doi: 10.1029/2009wr007706

Yamazaki D, Kanae S, Kim H, Oki T. 2011. A physically based description of floodplain inundation dynamics in a global river routing model. *Water Resources Research* **47**(4): W04501. doi: 10.1029/2010wr009726

Yan K, Di Baldassarre G, Solomatine DP. 2013. Exploring the potential of SRTM topographic data for flood inundation modelling under uncertainty. *Journal of hydroinformatics* **15**(3): 849. doi: 10.2166/hydro.2013.137

Yan K, Neal JC, Solomatine DP, Di Baldassarre, G. 2014. Global and low-cost topographic data to support flood studies. in *Hydro-meteorological Hazards, Risks, and Disasters*. 1 ed., edited by: Shroder JF, Paron P, Di Baldassarre G. Elsevier, 105-124, 2014.

Yan K, Tarpanelli A, Balint G, Moramarco T, Di Baldassarre G. 2015a. Exploring the Potential of SRTM Topography and Radar Altimetry to Support Flood Propagation Modeling: Danube Case Study. *Journal of Hydrologic Engineering* 20(2): 04014048.

Yan K, Di Baldassarre G, Solomatine DP, Schumann G. 2015b A review of low-cost space-borne data for hydraulic modelling: topography, flood extent and water level. *Hydrological Processes* doi:10.1002/hyp.10449

Yang L, Meng X, Zhang X. 2011. SRTM DEM and its application advances. *International Journal of Remote Sensing* **32**(14): 3875-3896.

Yamazaki D, O'Loughlin F, Trigg MA, Miller ZF, Pavelsky TM, Bates PD. 2014. Development of the Global Width Database for Large Rivers. *Water Resources Research* **50**(4): 3467-3480. doi: 10.1002/2013wr014664

Yin ZY, Wang X. 1999. A cross-scale comparison of drainage basin characteristics derived from digital elevation models. *Earth Surface Proceedings and Landform* **24**(6): 557–562.

Yu D, Lane SN. 2011. Interactions between subgrid-scale resolution, feature representation and grid-scale resolution in flood inundation modelling. *Hydrological Processes* **25**(1): 36-53. doi: 10.1002/hyp.7813

Yu D, Lane SN. 2006. Urban fluvial flood modelling using a two-dimensional diffusion-wave treatment, part 1: mesh resolution effects. *Hydrological Processes* **20**(7): 1541-1565. doi: 10.1002/hyp.5935

Zink M, TanDEM-X Mission Status. Proceedings of IGARSS (International Geoscience and Remote Sensing Symposium), Munich, Germany, July 22-27, 2012

## ACKNOWLEDGEMENTS

Everything starts from one afternoon in the summer of 2009, being deeply attracted by the beauty of water science, information technology and its perfect combination of hydroinformatics. I made my mind to come to the Netherlands and soon started my scientific journey. Now It is the time to thank someone who have helped and influnced me in the last five and half years.

First of all, I would like to express my deepest gratitude to Prof. Giuliano Di Baldassarre, who is my promotor as well as main supervisor for the last five years including my master thesis research. Thanks for involving me in KULTURisk Project and give me this wonderful oppotunity to unleash my scientific ensusiasm. I`ve leant a lot from him such as how to be an independent researcher with critical thinking; be ambitious while keep being productive; see things as opportunity and trun them into success; always hold an positive attitude towards challenges and issues. always get motivated and always, get ahead; Last but not the least, and perhaps one of the most important aspect, be confident and dare to dream.

I also would like to express my appreciation to my promotor Prof. Dimitri Solomatine. Your special sense of humor is definitely an asset in my PhD journey, while your rigorous attitude towards science has also influnced me significantly. I would also like to thank my journal paper co-authors who contribute to this thesis. Prof. Paul Bates, who provides me very efficient flood modelling tool together with strong techonical support. Dr. Jeffery Neal always resolve my endless query and deliver prompt solution regarding the modelling tool. Special thanks to Dr. Guy Schumann for his suggestions on remote sensing techniques in flood modelling, which not only helped me to finish the review paper, also enhanced my undestanding on this topic. Thanks Dr. Florian Pappenberger for his great support by providing ECMWF`s physical model chain.

I would like to thank Prof. Arthur Mynett and Dr. Ioana Popescu to let me take part in supervision and teaching activities for master programmes in UNESO-IHE. They boost my confidence and strengthen my various skills. In addition, I would like to thank all academic Staff members in UNESCO-IHE who have helped me in the last five years.

I would also like to extend my thanks to my colleague Micah Mukowel for his assistance in modelling, coding and for our happy (scientific) conversations during the PhD journey. It`s always nice to have a buddy like you to share thoughts. Big thanks to Maurizio Mazzoleni for all the funny jokes, relaxed coffee breaks, happy lunch time, and endless discussions. Thank the lunch group and colleagues in UNESCO-IHE (i.e. Juan Carlos Chacon, Blagoj Delipetrev, Alexander Jose Kaune, Pan Quan, Mario Castro, Gert Mulder, Oscar Marquez, Yared Abebe, Neiler Medina, Anuar Md. Ali, Patricia Trambauer, Jessica Salcedo, Veronica Minaya, Gonzalo Peña, Angelica Rada, Maria Reyes, Maribel Zapater) for the weekly relaxation Chinese/Italian restaurant and open market lunch tour. It was lovely to share all the sunny/windy/cloudy time with all of you! 'Ground floor (Oude Delft 95) Mafia' can always find excuse to enjoy nice party out of our intensive research life. This created a wonderful atmosphrere with great communication in both research and leisure time, helping each other to resolve all kinds of issues.

Special thanks to the various weekend dinner groups, hotpot club, philosophical talking, Sunday afternoon basketball madness, Saturday morning piano gathering (i.e. Xuan Zhu, Hui Chen, Taoping Wan, Qinghua Ye, Xiuhan Chen, Zheng Xu, Zhi Yang, Yuanyang Wan, Liqin Zuo, Shouqian Li, Shenglan Lu, Wei Song, Cong Li, Tingting Zhang, Nikola Stanic, Lihua Ma). It`s you who always keep me energetic and curious about everything in life.

I would like to thank my parents for your unconditional love and support. Thank my girlfriend Shanshan Li for your encouragement and accompany during the joyful and hard time of my PhD.


Kun Yan

Delft, the Netherlands

12 June 2015

## ABOUT THE AUTHOR

Kun Yan was born on 31st December 1985, in Bengbu, Anhui Province, China. He is the first child of the family with a 10-minutes younger twin brother. He grew up in this third-tier city of China with great expectations on his future, in the times that saw great changes, just like the vast majority of his single-child friends.

Luckily Kun entered Hohai University, a university of "211 Project" in China specialized in water resources. Kun received his bachelor degree from collage of Hydrology and Water Recourses, Hohai University on May, 2008. He then enrolled the master program of Ecohydrology in Hohai University. A year later, he moved to the Netherlands and joined the master program of Hydroinformatics of UNESCO-IHE. He started his PhD in UNESCO-IHE on July 2011, after two months obtained his MSc degree. His PhD topic is about the integration of low-cost space-borne data into hydraulic modelling of floods. Kun was also involved in the EC FP7 KULTURisk project, which aims at developing a culture of risk prevention for nature disasters including floods.

## Publications

**Peer-reviewed Journal Papers**

**K. Yan**\*, G. Di Baldassarre, D.P. Solomatine, F. Pappenberger 2015 Flood mapping in data-scarce areas: regional versus physically-based approaches for design flood estimation. *Hydrological Sciences Journal*. Under review

**K. Yan**\*, G. Di Baldassarre, D.P. Solomatine, G. J-P. Schumann 2015 A review of low-cost space-borne data for hydraulic modelling: topography, flood extent and water level. *Hydrological Processes*. doi:10.1002/hyp.10449

**K. Yan**\*, A. Tarpanelli, G. Balint, T. Moramarco, G. Di Baldassarre 2015 Exploring the Potential of Radar Altimetry and SRTM Topography to Support Flood Propagation Modelling: the Danube Case Study. *Journal of Hydrologic Engineering*. 20(2), 04014048. doi:10.1061/(ASCE)HE.1943-5584.0001018

**K. Yan**\*, G. Di Baldassarre, D.P. Solomatine 2013 Exploring the potential of SRTM topographic data for flood inundation modelling under uncertainty. *Journal of Hydroinformatics*. 15(3), 849–861. doi: 10.2166/hydro.2013.137

M. Mukolwe, **K. Yan**\*, G. Di Baldassarre, D.P. Solomatine 2015 Testing new sources of topographic data for flood propagation modelling under structural, parameter and observation uncertainty. *Hydrological Sciences Journal*. doi: 10.1080/02626667.2015.1019507

G. Di Baldassarre*, A. Viglione, G. Carr, L. Kuil, **K. Yan**, L. Brandimarte and G. Blöschl 2015 Debates—Perspectives on sociohydrology: Capturing feedbacks between physical and social processes. *Water Resources Research.* DOI: 10.1002/2014WR016416

G. Di Baldassarre*, **K. Yan**, R. Ferdous, L. Brandimarte 2014 The interplay between human population dynamics and flooding in Bangladesh: a spatial analysis. Proc. IAHS, 364, 188-191, doi:10.5194/piahs-364-188-2014, 2014.


**Book Chapters**

**K. Yan**\*, J. Neal, D.P. Solomatine, G. Di Baldassarre 2014 Global and low-cost topographic data to support flood studies, in: *Hydro-Meteological Hazards, Risks, and Disasters.* 1 ed., edited by: Shroder, J. F., Paron, P., and Di Baldassarre, G., Elsevier, 105-124, 2014. doi:10.1016/B978-0-12-394846-5.00004-7


**Conference Proceedings**

K. Yan, F. Pappenberger, Y. M. Umer, D. P. Solomatine, G. Di Baldassarre 2014 Regional versus physically-based methods for flood inundation modelling in data scarce areas: an application to the Blue Nile. *Proceedings of the 11th International Conference on Hydroinformatics (HIC 2014)*, New York, USA.

G. Di Baldassarre, G. Schumann, D. Solomatine, K. Yan, and P.D. Bates. 2012. Global flood mapping: current issues and future directions. *Proceedings of the 10th International Conference on Hydroinformatics (HIC 2012)*, Hamburg, Germany.

E. Ridolfi, K. Yan, L. Alfonso, G. Di Baldassarre, F. Napolitano, F. Russo, and P.D. Bates. 2012 An entropy method for floodplain monitoring network design. *Numerical Analysis and Applied Mathematics ICNAAM 2012 AIP Conference Proceedings.* 1479, 1780-1783 (2012), Kos, Greece, doi: 10.1063/1.4756522


**Conference Contributions**

G. Di Baldassarre, A. Viglione, K. Yan, L. Brandimarte, and G. Blöschl. 2014 Long-term dynamics emerging in floodplains and deltas from the interactions between hydrology and society in a changing climate. EGU General Assembly 2014, Geophysical Research Abstracts, Vol. 16, EGU2014-1948, 2014

K. Yan, G. Di Baldassarre, F. Pappenberger, and D. Solomatine. 2014 Flood inundation modelling in data-poor areas: a case study. EGU General Assembly 2014, Geophysical Research Abstracts, Vol. 16, EGU2014-16556, 2014

F Nardi, K Yan, G Di Baldassarre, S Grimaldi 2013 Large scale floodplain mapping using a hydrogeomorphic method AGU Fall Meeting Abstracts, Vol. 1, 2013

K. Yan, G. Di Baldassarre, D.P. Solomatine. 2013 Can we use SRTM topographic data for large river flood modelling under uncertainty? EGU 2013 Leonardo Conference, 17-21 October, Kos, Greece.

K. Yan, G. Di Baldassarre, J. Neal, D. P. Solomatine. 2013 Flood imagery, hydrometric data and downstream water surface slope for constraining uncertainty in inundation modelling based on SRTM and LiDAR topography. EGU General Assembly 2013, Geophysical Research Abstracts, 15, 2013

K. Yan., Di Baldassarre, G. and Solomatine, D.P. Exploring the usefulness of global topography to support flood management under uncertainty. EGU 2012 Leonardo Conference, Turin, Italy.

K. Yan., Di Baldassarre, G. and Solomatine, D.P. Assessing the usefulness of SRTM topography to support hydraulic modelling under uncertainty. IAHS-PUB Symposium 2012, Delft, the Netherlands.

K. Yan., Di Baldassarre, G., Solomatine, D.P. & Giustarini, L. Flood Inundation Modelling Under Uncertainty Using Globally and Freely Available Remote Sensing Data. EGU General Assembly 2012, Geophysical Research Abstracts, 14, 2012

E. Ridolfi, K. Yan., Alfonso, L., Di Baldassarre, G., Napolitano, F., Russo, F. & Bates, P.D. Optimization of floodplain monitoring sensors through an entropy approach. EGU General Assembly 2012, Geophysical Research Abstracts, 14, 2012

G. Di Baldassarre, Schumann, G., Solomatine, D., K. Yan., and Bates, P.D. 2011. Can we map floodplains globally? Geophysical Research Abstracts Vol.13, EGU2011-5694-1.

K. Yan., Di Baldassarre, G., Solomatine, D.P. & Dottori, F. Flood Inundation Modelling in Large Rivers Under Uncertainty Using Globally and Freely Available Remote Sensing Data. EGU 2011 Leonardo Conference, 23-25 November, Bratislava, Slovakia.

T - #0433 - 101024 - C32 - 240/170/7 - PB - 9781138028753 - Gloss Lamination